Multilinear algebra

Multilinear algebra

D. G. NORTHCOTT F.R.S.
Formerly Town Trust Professor of Mathematics at the University of Sheffield

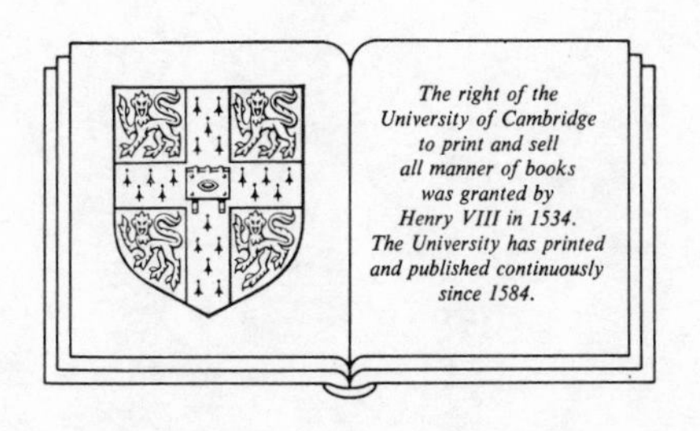

CAMBRIDGE UNIVERSITY PRESS
Cambridge
London New York New Rochelle
Melbourne Sydney

Published by the Press Syndicate of the University of Cambridge
The Pitt Building, Trumpington Street, Cambridge CB2 1RP
32 East 57th Street, New York, NY 10022, USA
296 Beaconsfield Parade, Middle Park, Melbourne 3206, Australia

First published 1984

Printed in Great Britain at the University Press, Cambridge

Library of Congress catalogue card number: 83-27210

British Library cataloguing in publication data
Northcott, D.G.
Multilinear algebra.
1. Algebras, Linear
I. Title
512'.5 QA184

ISBN 0 521 26269 0

MP

Contents

Preface

This account of Multilinear Algebra has developed out of lectures which I gave at the University of Sheffield during the session 1981/2. In its present form it is designed for advanced undergraduates and those about to commence postgraduate studies. At this general level the only special prerequisite for reading the whole book is a familiarity with the notion of a module (over a commutative ring) and with such concepts as submodule, factor module and homomorphism.

Multilinear Algebra arises out of Linear Algebra and like its antecedent is a subject which has applications in a great many different fields. Indeed, there are so many reasons why mathematicians may need some knowledge of its concepts and results that any selection of applications is likely to disappoint as many readers as it satisfies. Furthermore, such a selection tends to upset the balance of the subject as well as adding substantially to the required background knowledge. It is my impression that young mathematicians often acquire their knowledge of Multilinear Algebra in a rather haphazard and fragmentary fashion. Here I have attempted to weld the most commonly used fragments together and to fill out the result so as to obtain a theory with an easily recognizable structure.

The book begins with the study of multilinear mappings and the tensor, exterior and symmetric powers of a module. Next, the tensor powers are fitted together to produce the tensor algebra of a module, and a similar procedure yields the exterior and symmetric algebras. Multilinear mappings and the three algebras just mentioned form the most widely used parts of the subject and, in this account, occupy the first six chapters. However, at this point we are at the threshold of a richer theory, and it is Chapter 7 that provides the climax of the book.

Chapter 7 starts with the observation that if we re-define algebras in terms of certain commutative diagrams, then we are led to a dual concept known as a *coalgebra.* Now it sometimes happens that, on the same underlying set, there exist simultaneously both an algebra-structure and a coalgebra-structure. When this happens, and provided that the two structures interact suitably, the result is called a *Hopf algebra.* It turns out that exterior and symmetric algebras are better regarded as Hopf algebras.

This approach confers further benefits. By considering linear forms on a coalgebra it is always possible to construct an associated algebra; and, since exterior and symmetric algebras have a coalgebra-structure, this construction may be applied to them. The result in the first case is the algebra of differential forms (the Grassmann algebra) and in the second case it is the algebra of differential operators.

The final chapter deals with *graded duality.* From every graded module we can construct another graded module known as its graded dual. If the components of the original graded module are free and of finite rank, then this process, when applied twice, yields a double dual that is a copy of the graded module with which we started. For similarly restricted graded algebras, coalgebras and Hopf algebras this technique gives rise to a full duality; algebras become coalgebras and *vice versa*; and Hopf algebras continue to be Hopf algebras.

Each chapter has, towards its end, a section with the title 'Comments and exercises'. The comments serve to amplify the main theory and to draw attention to points that require special attention; the exercises give the reader an opportunity to test his or her understanding of the text and a chance to become acquainted with additional results. Some exercises are marked with an asterisk. Usually these exercises are selected on the grounds of being particularly interesting or more than averagely difficult; sometimes they contain results that are used later. Where an asterisk is attached to an exercise a solution is provided in the following section. However, to prevent gaps occurring in the argument, a result contained in an exercise is not used later unless a solution has been supplied.

Once the guide-lines for the book had been settled, I found that the subject unfolded very much under its own momentum. Where I had to consult other sources, I found C. Chevalley's *Fundamental Concepts of Algebra*, even though it was written more than a quarter of a century ago, especially helpful. In particular, the account given here of Pfaffians follows closely that given by Chevalley.

Finally I wish to record my thanks to Mrs E. Benson and Mrs J. Williams of the Department of Pure Mathematics at Sheffield University. Between them they typed the whole book; and their cheerful co-operation enabled the exacting task of preparing it for the printers to proceed smoothly and without a hitch.

Sheffield, April 1983 D. G. Northcott

1

Multilinear mappings

General remarks

Throughout Chapter 1 the letter R will denote a *commutative* ring which possesses an identity element. R is called *trivial* if its zero element and its identity element are the same. Of course if R is trivial, then all its modules are null modules.

The standard notation for tensor products is introduced in Section (1.2) and from there on we allow ourselves the freedom (in certain contexts) to omit the suffix which indicates the ring over which the products are formed. More precisely when (in Chapter 1) we are dealing with tensor products of R-modules, we sometimes use $\otimes$ rather than the more explicit $\otimes_R$. This is done solely to avoid typographical complications.

1.1 Multilinear mappings

Let $M_1, M_2, \ldots, M_p$ $(p \geq 1)$ and M be R-modules and let

$$\phi: M_1 \times M_2 \times \cdots \times M_p \to M \tag{1.1.1}$$

be a mapping of the cartesian product $M_1 \times M_2 \times \cdots \times M_p$ into M. We use $m_1, m_2, \ldots, m_p$ to denote typical elements of $M_1, M_2, \ldots, M_p$ respectively and r to denote a typical element of R. The mapping ϕ is called *multilinear* if

$$\phi(m_1, \ldots, m_i' + m_i'', \ldots, m_p) = \phi(m_1, \ldots, m_i', \ldots, m_p) + \phi(m_1, \ldots, m_i'', \ldots, m_p) \tag{1.1.2}$$

and

$$\phi(m_1, \ldots, rm_i, \ldots, m_p) = r\phi(m_1, \ldots, m_i, \ldots, m_p). \tag{1.1.3}$$

(Here, of course, i is unrestricted provided it lies between 1 and p.) For example, when $p = 1$ a multilinear mapping is the same as a homomorphism of R-modules.

Suppose now that (1.1.1) is a multilinear mapping. We can derive other multilinear mappings from it in the following way. Let $h: M \to N$ be a homomorphism of R-modules. Then $h \circ \phi$ is a multilinear mapping of $M_1 \times M_2 \times \cdots \times M_p$ into N. This raises the question as to whether it is possible to choose M and ϕ so that every multilinear mapping of $M_1 \times M_2 \times \cdots \times M_p$ can be obtained in this way. More precisely we pose

Problem 1. *To choose M and ϕ in such a way that given any multilinear mapping*

$$\psi: M_1 \times M_2 \times \cdots \times M_p \to N$$

there is exactly one homomorphism $h: M \to N$ (of R-modules) such that $h \circ \phi = \psi$.

This will be referred to as the *universal problem* for the multilinear mappings of $M_1 \times M_2 \times \cdots \times M_p$.

We begin by observing that if the pair (M, ϕ) solves the universal problem, then whenever we have homomorphisms $h_i: M \to N$ $(i=1, 2)$ such that $h_1 \circ \phi = h_2 \circ \phi$, then necessarily $h_1 = h_2$. Now suppose that (M, ϕ) and (M', ϕ') both solve our universal problem. In this situation there will exist unique R-homomorphisms $\lambda: M \to M'$ and $\lambda': M' \to M$ such that $\lambda \circ \phi = \phi'$ and $\lambda' \circ \phi' = \phi$. It follows that $(\lambda' \circ \lambda) \circ \phi = \phi$ or $(\lambda' \circ \lambda) \circ \phi = i \circ \phi$, where i is the identity mapping of M. The observation at the beginning of this paragraph now shows that $\lambda' \circ \lambda = i$ and similarly $\lambda \circ \lambda'$ is the identity mapping of M'. But this means that $\lambda: M \to M'$ and $\lambda': M' \to M$ are inverse isomorphisms. Thus if we have two solutions of the universal problem, then they will be copies of each other in a very precise sense. More informally we may say that *if Problem 1 has a solution, then the solution is essentially unique.* Before we consider whether a solution always exists, we make some general observations about free modules.

From here on, until we come to the statement of Theorem 1, we shall assume that R is *non-trivial.* We recall that an R-module which possesses a *linearly independent* system of generators is called *free* and a linearly independent system of generators of a free module is called a *base.* Now let X be a set and consider homogeneous linear polynomials (with coefficients in R) in the elements of X. These form a free R-module having the elements of X as a base. This is known as the *free module generated by* X. (If X is empty, then the free module which it generates is the null module.) Any mapping of X into an R-module N has exactly one extension to an R-homomorphism of this free module into N.

We are now ready to solve Problem 1. Let $U(M_1, M_2, \ldots, M_p)$ be the free R-module generated by the cartesian product $M_1 \times M_2 \times \cdots \times M_p$. Of course, this has the set of sequences $(m_1, m_2, \ldots, m_p)$ as a base. The elements of $U(M_1, M_2, \ldots, M_p)$ that have one or other of the forms

$$(m_1, \ldots, m_i' + m_i'', \ldots, m_p) - (m_1, \ldots, m_i', \ldots, m_p) \\ - (m_1, \ldots, m_i'', \ldots, m_p) \quad (1.1.4)$$

and

$$(m_1, \ldots, rm_i, \ldots, m_p) - r(m_1, \ldots, m_i, \ldots, m_p) \quad (1.1.5)$$

generate a submodule $V(M_1, M_2, \ldots, M_p)$ say. Put

$$M = U(M_1, M_2, \ldots, M_p)/V(M_1, M_2, \ldots, M_p) \quad (1.1.6)$$

and define

$$\phi: M_1 \times M_2 \times \cdots \times M_p \to M \quad (1.1.7)$$

so that $\phi(m_1, m_2, \ldots, m_p)$ is the natural image of $(m_1, m_2, \ldots, m_p)$, considered as an element of $U(M_1, M_2, \ldots, M_p)$, in M. Since the elements of $U(M_1, M_2, \ldots, M_p)$ described in (1.1.4) and (1.1.5) become zero in M, ϕ satisfies (1.1.2) and (1.1.3) and therefore it is multilinear. It will now be shown that M and ϕ provide a solution to our universal problem.

To this end suppose that

$$\psi: M_1 \times M_2 \times \cdots \times M_p \to N$$

is multilinear. There is an R-homomorphism

$$U(M_1, M_2, \ldots, M_p) \to N$$

in which $(m_1, m_2, \ldots, m_p)$ is mapped into $\psi(m_1, m_2, \ldots, m_p)$. Since ψ is multilinear, the homomorphism maps the elements (1.1.4) and (1.1.5) into zero and therefore it vanishes on $V(M_1, M_2, \ldots, M_p)$. Accordingly there is induced a homomorphism $h: M \to N$ which satisfies $h(\phi(m_1, m_2, \ldots, m_p)) = \psi(m_1, m_2, \ldots, m_p)$. Thus $h \circ \phi = \psi$. Finally if $h': M \to N$ is also a homomorphism such that $h' \circ \phi = \psi$, then h and h' have the same effect on every element of the form $\phi(m_1, m_2, \ldots, m_p)$. However, these elements generate M as an R-module and therefore $h = h'$. Thus (M, ϕ) solves the universal problem. We sum up our results so far.

Theorem 1. *Let $M_1, M_2, \ldots, M_p$ $(p \geq 1)$ be R-modules. Then the universal problem for multilinear mappings of $M_1 \times M_2 \times \cdots \times M_p$ has a solution. Furthermore the solution is essentially unique (in the sense explained previously).*

Corollary. *Suppose that (M, ϕ) solves the universal problem described above. Then each element of M can be expressed as a finite sum of elements of the form $\phi(m_1, m_2, \ldots, m_p)$.*

Proof. A solution to Problem 1 has just been constructed and it would be easy to check that this particular solution has the property described in the corollary. We could then utilize the result which says that any two solutions are virtually identical. However, it is more interesting to base a proof directly on the fact that (M, ϕ) meets the requirements of the universal problem. This is the method employed here.

Suppose then that M' is the R-submodule of M generated by elements of the form $\phi(m_1, m_2, \ldots, m_p)$. Also let $h_1 : M \to M/M'$ be the natural homomorphism and $h_2 : M \to M/M'$ the null homomorphism. Then $h_1 \circ \phi = h_2 \circ \phi$ and therefore $h_1 = h_2$. However, this implies that $M = M'$.

Let $x \in M = M'$. Then x can be expressed in the form

$$x = r\phi(m_1, m_2, \ldots, m_p) + r'\phi(m'_1, m'_2, \ldots, m'_p) + \cdots,$$

where the sum is finite. However,

$$r\phi(m_1, m_2, \ldots, m_p) = \phi(rm_1, m_2, \ldots, m_p)$$

and similarly in the case of the other terms. The corollary follows.

1.2 The tensor notation

Once again $M_1, M_2, \ldots, M_p$, where $p \geq 1$, denote R-modules and we continue to use $m_1, m_2, \ldots, m_p$ to denote typical elements of these modules, and r to denote a typical element of R. Let the pair (M, ϕ) provide a solution to the universal problem for multilinear mappings of $M_1 \times M_2 \times \cdots \times M_p$. It is customary to write

$$M = M_1 \otimes_R M_2 \otimes_R \cdots \otimes_R M_p \tag{1.2.1}$$

and to use $m_1 \otimes m_2 \otimes \cdots \otimes m_p$ to designate the element $\phi(m_1, m_2, \ldots, m_p)$ of M. When this notation is employed, $M_1 \otimes_R M_2 \otimes_R \cdots \otimes_R M_p$ is called the *tensor product* of $M_1, M_2, \ldots, M_p$ over R.

It will be recalled that the solution to the universal problem is unique only to the extent that any two solutions are copies of each other. Thus we can have different *models* for the tensor product. However, if we have two such models, then they are isomorphic (as modules) under an isomorphism which matches the element represented by $m_1 \otimes m_2 \otimes \cdots \otimes m_p$ in the first model with the similarly represented element in the second. On account of this we usually do not need to specify which particular model we are using.

Next, because ϕ is multilinear, the relations

$$\begin{aligned} m_1 \otimes \cdots \otimes (m'_i + m''_i) \otimes \cdots \otimes m_p \\ = m_1 \otimes \cdots \otimes m'_i \otimes \cdots \otimes m_p + m_1 \otimes \cdots \otimes m''_i \otimes \cdots \otimes m_p \end{aligned} \tag{1.2.2}$$

and

$$m_1 \otimes \cdots \otimes rm_i \otimes \cdots \otimes m_p = r(m_1 \otimes \cdots \otimes m_i \otimes \cdots \otimes m_p) \tag{1.2.3}$$

both hold. Moreover the corollary to Theorem 1 shows that each element of $M_1 \otimes_R M_2 \otimes_R \cdots \otimes_R M_p$ is a finite sum of elements of the form $m_1 \otimes m_2 \otimes \cdots \otimes m_p$. Finally the fact that the tensor product provides the solution to the universal problem for multilinear mappings may be restated as

Theorem 2. *Given an R-module N and a multilinear mapping*

$$\psi: M_1 \times M_2 \times \cdots \times M_p \to N$$

there exists a unique R-module homomorphism h, of $M_1 \otimes_R M_2 \otimes_R \cdots \otimes_R M_p$ *into N, such that*

$$h(m_1 \otimes m_2 \otimes \cdots \otimes m_p) = \psi(m_1, m_2, \ldots, m_p)$$

for all $m_1, m_2, \ldots, m_p$.

We interrupt the main argument to observe that when $p=1$ the universal problem can be solved by taking M to be M_1 and ϕ to be the identity mapping. This confirms that for $p=1$ the tensor product $M_1 \otimes_R M_2 \otimes_R \cdots \otimes_R M_p$ is just M_1 as we should naturally expect.

The reader will have noticed that the full tensor notation is rather heavy. To counteract this we shall often use a simplified version. Indeed, because in Chapter 1 we shall only be concerned with a single ring R, it will not cause confusion if we use $M_1 \otimes M_2 \otimes \cdots \otimes M_p$ in place of the typographically more cumbersome but more explicit $M_1 \otimes_R M_2 \otimes_R \cdots \otimes_R M_p$. Nevertheless, in the statement of theorems and other results likely to be referred to later we shall restore the subscript which identifies the relevant ring.

So much for matters of notation. Before we leave this section we shall seek to gain insight into the nature of tensor products by examining the result of forming the tensor product of a number of *free* modules.

Suppose then that, for $1 \le i \le p$, M_i is a free R-module and that B_i is a base for M_i. The first point to note is that any mapping of $B_1 \times B_2 \times \cdots \times B_p$ into an R-module N has precisely one extension to a multilinear mapping of $M_1 \times M_2 \times \cdots \times M_p$ into N. We use this observation in the proof of our next theorem.

Theorem 3. *Let* M_i $(i=1, 2, \ldots, p)$ *be a free R-module with a base* B_i. *Then* $M_1 \otimes_R M_2 \otimes_R \cdots \otimes_R M_p$ *is also a free R-module and it has the elements* $b_1 \otimes b_2 \otimes \cdots \otimes b_p$, *where* $b_i \in B_i$, *as a base.*

Proof. Denote by M the free R-module generated by the set $B_1 \times B_2 \times \cdots \times B_p$. Thus the sequences $(b_1, b_2, \ldots, b_p)$ form a base for M. There is then a mapping ϕ, of $B_1 \times B_2 \times \cdots \times B_p$ into M, in which $\phi(b_1, b_2, \ldots, b_p)$ is the base element $(b_1, b_2, \ldots, b_p)$. Now, as we noted above, the mapping has an extension (denoted by the same letter) to a multilinear mapping of $M_1 \times M_2 \times \cdots \times M_p$ into M. Clearly all we need to do to complete the proof is to show that (M, ϕ) solves Problem 1.

Suppose then that we have a multilinear mapping

$$\psi: M_1 \times M_2 \times \cdots \times M_p \to N.$$

In these circumstances there exists an R-homomorphism $h: M \to N$ which maps $(b_1, b_2, \ldots, b_p)$ into $\psi(b_1, b_2, \ldots, b_p)$. Obviously $h \circ \phi = \psi$. Moreover, if $h': M \to N$ is also a homomorphism satisfying $h' \circ \phi = \psi$, then h and h' agree on the base $B_1 \times B_2 \times \cdots \times B_p$ of M and therefore $h = h'$. The proof is therefore complete.

1.3 Tensor powers of a module

Let M be an R-module and $p \geq 1$ an integer. Put

$$T_p(M) = M \otimes_R M \otimes_R \cdots \otimes_R M, \tag{1.3.1}$$

where there are p factors. The R-module $T_p(M)$ is called the p-th *tensor power* of M. These powers will later form the components of a graded algebra known as the *tensor algebra* of M. For the moment we note that $T_1(M) = M$. Also, if M is a free R-module and B is a base of M, then $T_p(M)$ is also free and it has the elements $b_1 \otimes b_2 \otimes \cdots \otimes b_p$, where $b_i \in B$, as a base. This follows from Theorem 3.

1.4 Alternating multilinear mappings

As in the last section M denotes an R-module. In this section we shall have occasion to consider products such as $M \times M \times \cdots \times M$ and $M \otimes M \otimes \cdots \otimes M$. Whenever such a product occurs it is to be understood that the number of factors is p, where $p \geq 1$.

A multilinear mapping

$$\eta: M \times M \times \cdots \times M \to N$$

is called *alternating* if $\eta(m_1, m_2, \ldots, m_p) = 0$ whenever the sequence $(m_1, m_2, \ldots, m_p)$ contains a repetition. (To clarify the position when $p = 1$ we make it explicit that all linear mappings $M \to N$ are regarded as alternating.)

Suppose that η has this property. If $1 \leq i < j \leq p$, then on expanding

$$\eta(m_1, \ldots, m_i + m_j, \ldots, m_i + m_j, \ldots, m_p) = 0,$$

where the element $m_i + m_j$ occurs in both the i-th and j-th positions, we find that

$$\eta(m_1, \ldots, m_i, \ldots, m_j, \ldots, m_p) + \eta(m_1, \ldots, m_j, \ldots, m_i, \ldots, m_p) = 0.$$

Thus if we interchange two terms in the sequence $(m_1, m_2, \ldots, m_p)$ the effect on $\eta(m_1, m_2, \ldots, m_p)$ is to multiply it by -1. From this observation we at once obtain

Lemma 1. *Let* $\eta: M \times M \times \cdots \times M \to N$ *be an alternating multilinear mapping and let* $(i_1, i_2, \ldots, i_p)$ *be a permutation of* $(1, 2, \ldots, p)$. *Then*

$$\eta(m_{i_1}, m_{i_2}, \ldots, m_{i_p}) = \pm\eta(m_1, m_2, \ldots, m_p),$$

where the sign is plus if the permutation is even and minus if it is odd.

Another useful observation is recorded in

Lemma 2. *Let* $\eta: M \times M \times \cdots \times M \to N$ *be a multilinear mapping and suppose that* $\eta(m_1, m_2, \ldots, m_p) = 0$ *whenever* $m_i = m_{i+1}$ *for some i. Then* η *is an alternating mapping.*

Proof. The argument just before the statement of Lemma 1 shows that $\eta(m_1, m_2, \ldots, m_p)$ changes sign if we interchange two *adjacent* terms in the sequence $(m_1, m_2, \ldots, m_p)$. Now suppose that $(m_1, m_2, \ldots, m_p)$ contains a repetition. Then either two equal terms occur next to each other or this situation can be brought about by a number of adjacent interchanges. It follows that $\eta(m_1, m_2, \ldots, m_p) = 0$ so the lemma is proved.

Consider an alternating multilinear mapping η, of $M \times M \times \cdots \times M$ into an R-module N, and suppose that $h: N \to K$ is a homomorphism of R-modules. Then $h \circ \eta$ is an alternating multilinear mapping of $M \times M \times \cdots \times M$ into K. This observation leads us to pose the following universal problem.

Problem 2. *To choose N and* η *in such a way that given any alternating multilinear mapping*

$$\zeta: M \times M \times \cdots \times M \to K$$

there exists exactly one homomorphism $h: N \to K$ (*of R-modules*) *such that* $\zeta = h \circ \eta$.

To solve this problem we consider the tensor power $T_p(M)$ and denote by $J_p(M)$ the submodule generated by all elements $m_1 \otimes m_2 \otimes \cdots \otimes m_p$, where $(m_1, m_2, \ldots, m_p)$ contains a repetition. (It is understood that $J_1(M) = 0$.) Put $N = T_p(M)/J_p(M)$ and let η be the mapping of $M \times M \times \cdots \times M$ into N which takes $(m_1, m_2, \ldots, m_p)$ into the natural image of $m_1 \otimes m_2 \otimes \cdots \otimes m_p$ in N. Then η is an alternating multilinear mapping. If now

$$\zeta: M \times M \times \cdots \times M \to K$$

is also an alternating multilinear mapping, then, by Theorem 2, there is a homomorphism of $M \otimes M \otimes \cdots \otimes M$ into K which takes $m_1 \otimes m_2 \otimes \cdots \otimes m_p$ into $\zeta(m_1, m_2, \ldots, m_p)$. This homomorphism vanishes on $J_p(M)$ and so it induces a homomorphism $h: N \to K$. It is clear that $h \circ \eta = \zeta$. If we have a second homomorphism, say $h': N \to K$, and this satisfies $h' \circ \eta = \zeta$, then h and h' agree on the elements $\eta(m_1, m_2, \ldots, m_p)$ and therefore they agree on a system of generators of N. But this ensures that $h = h'$.

It has now been shown that the universal problem for alternating multilinear mappings of $M \times M \times \cdots \times M$ has a solution. The solution is unique in the following sense. Suppose that (N, η) and (N', η') both solve Problem 2. Then there are inverse isomorphisms $\lambda: N \to N'$ and $\lambda': N' \to N$ such that $\lambda \circ \eta = \eta'$ and $\lambda' \circ \eta' = \eta$. The situation is, in fact, almost identical with that encountered in dealing with uniqueness in the case of Problem 1.

Let us suppose that (N, η) solves the universal problem for alternating multilinear mappings of $M \times M \times \cdots \times M$. Put

$$E_p(M) = N \tag{1.4.1}$$

and

$$m_1 \wedge m_2 \wedge \cdots \wedge m_p = \eta(m_1, m_2, \ldots, m_p). \tag{1.4.2}$$

Then, because η is multilinear, we have

$$\begin{aligned} &m_1 \wedge \cdots \wedge (m_i' + m_i'') \wedge \cdots \wedge m_p \\ &\qquad = m_1 \wedge \cdots \wedge m_i' \wedge \cdots \wedge m_p + m_1 \wedge \cdots \wedge m_i'' \wedge \cdots \wedge m_p \end{aligned} \tag{1.4.3}$$

and

$$m_1 \wedge \cdots \wedge rm_i \wedge \cdots \wedge m_p = r(m_1 \wedge \cdots \wedge m_i \wedge \cdots \wedge m_p). \tag{1.4.4}$$

But η is also alternating. Consequently $m_1 \wedge m_2 \wedge \cdots \wedge m_p = 0$ whenever $(m_1, m_2, \ldots, m_p)$ contains a repetition and, by Lemma 1,

$$m_{i_1} \wedge m_{i_2} \wedge \cdots \wedge m_{i_p} = \pm m_1 \wedge m_2 \wedge \cdots \wedge m_p, \tag{1.4.5}$$

where the plus sign is to be used if $(i_1, i_2, \ldots, i_p)$ is an even permutation of $(1, 2, \ldots, p)$ and the minus sign if the permutation is odd.

The module $E_p(M)$ is called the p-th *exterior power* of M. As we have seen it is unique in much the same sense that $T_p(M)$ is unique. Since when $p=1$ we can solve Problem 2 by means of M and its identity mapping, we have $E_1(M) = M$.

The defining property of the exterior power is restated in the next theorem.

Theorem 4. *Given an R-module K and an alternating multilinear mapping*

$$\zeta: M \times M \times \cdots \times M \to K$$

there exists a unique R-homomorphism $h: E_p(M) \to K$ *such that*

$$h(m_1 \wedge m_2 \wedge \cdots \wedge m_p) = \zeta(m_1, m_2, \ldots, m_p)$$

for all $m_1, m_2, \ldots, m_p$ *in* M.

Corollary. *Each element of* $E_p(M)$ *is a finite sum of elements of the form* $m_1 \wedge m_2 \wedge \cdots \wedge m_p$.

A proof can be obtained by means of a trivial modification of the argument used to establish the corollary to Theorem 1, so details will not be given.

The mapping of $M \times M \times \cdots \times M$ into $E_p(M)$ which takes $(m_1, m_2, \ldots, m_p)$ into $m_1 \wedge m_2 \wedge \cdots \wedge m_p$ is multilinear and so, by Theorem 2, there is induced a homomorphism $T_p(M) \to E_p(M)$. This is surjective because the image of $m_1 \otimes m_2 \otimes \cdots \otimes m_p$ is $m_1 \wedge m_2 \wedge \cdots \wedge m_p$. We refer to $T_p(M) \to E_p(M)$ as the *canonical homomorphism* of the tensor power onto the exterior power. Note that when $p=1$ the canonical homomorphism is the identity mapping of M if we make the identifications $T_1(M) = M = E_1(M)$.

As before we let $J_p(M)$ denote the R-submodule of $T_p(M)$ generated by all products $m_1 \otimes m_2 \otimes \cdots \otimes m_p$ in which there is a repeated factor.

Theorem 5. *The canonical homomorphism $T_p(M) \to E_p(M)$ is surjective and its kernel is $J_p(M)$. Moreover $J_p(M)$ is generated, as an R-module, by all products $m_1 \otimes m_2 \otimes \cdots \otimes m_p$, where $m_i = m_{i+1}$ for some i.*

Proof. Let $J'_p(M)$ be the submodule of $T_p(M)$ generated by all $m_1 \otimes m_2 \otimes \cdots \otimes m_p$ with $m_i = m_{i+1}$ for some i, and let $J''_p(M)$ be the kernel of the canonical homomorphism. Then $J'_p(M) \subseteq J_p(M) \subseteq J''_p(M)$. Next, by Lemma 2, the mapping

$$\theta: M \times M \times \cdots \times M \to T_p(M)/J'_p(M),$$

in which $\theta(m_1, m_2, \ldots, m_p) = m_1 \otimes m_2 \otimes \cdots \otimes m_p + J'_p(M)$, is multilinear and alternating and therefore, by Theorem 4, there is a homomorphism $\lambda: E_p(M) \to T_p(M)/J'_p(M)$ such that

$$\lambda(m_1 \wedge m_2 \wedge \cdots \wedge m_p) = m_1 \otimes m_2 \otimes \cdots \otimes m_p + J'_p(M).$$

But if λ is combined with the canonical homomorphism $T_p(M) \to E_p(M)$, then the result is the natural homomorphism

$$T_p(M) \to T_p(M)/J'_p(M).$$

Thus the natural homomorphism vanishes on $J''_p(M)$ and therefore $J''_p(M) \subseteq J'_p(M)$. Accordingly $J'_p(M) = J_p(M) = J''_p(M)$ and the theorem is proved.

It is useful to know the structure of the exterior powers of a free module. Let us assume therefore that M is a free R-module and that B is one of its bases. On the set formed by all sequences $(b_1, b_2, \ldots, b_p)$ of p *distinct* elements of B we introduce the equivalence relation in which $(b_1, b_2, \ldots, b_p)$ and $(b'_1, b'_2, \ldots, b'_p)$ are regarded as equivalent if each is a permutation of the other. From each equivalence class we now select a single representative. In the next theorem the set formed by these representatives is denoted by $I_p(B)$.

Theorem 6. *Let M be a free R-module having the set B as a base. Then $E_p(M)$ is a free R-module having the elements $b_1 \wedge b_2 \wedge \cdots \wedge b_p$ as a base, where $(b_1, b_2, \ldots, b_p)$ ranges over $I_p(B)$.*

Proof. Let N be the free R-module generated by $I_p(B)$. Define a mapping

$$\eta: B \times B \times \cdots \times B \to N$$

as follows. When $(b_1, b_2, \ldots, b_p)$ contains a repetition $\eta(b_1, b_2, \ldots, b_p)$ is to be zero. When, however, $b_1, b_2, \ldots, b_p$ are all different there is a unique permutation $(b'_1, b'_2, \ldots, b'_p)$, of $(b_1, b_2, \ldots, b_p)$, such that $(b'_1, b'_2, \ldots, b'_p)$ belongs to $I_p(B)$; in this case we put

$$\eta(b_1, b_2, \ldots, b_p) = \pm(b'_1, b'_2, \ldots, b'_p),$$

where the plus sign is taken if the permutation is even and the minus sign if it is odd. The mapping η has a unique extension (denoted by the same letter) to a multilinear mapping of $M \times M \times \cdots \times M$ into N, and the construction ensures that the extension is alternating as well as multilinear.

We claim that (N, η) solves Problem 2. Clearly once this has been established the theorem will follow. Suppose then that

$$\zeta: M \times M \times \cdots \times M \to K$$

is an alternating multilinear mapping. Since N is free, there is a homomorphism $h: N \to K$ such that $h(b_1, b_2, \ldots, b_p) = \zeta(b_1, b_2, \ldots, b_p)$ whenever $(b_1, b_2, \ldots, b_p)$ is in $I_p(B)$. Thus $h \circ \eta$ and ζ are alternating multilinear mappings which agree on $I_p(B)$ and therefore on $B \times B \times \cdots \times B$ as well. If follows that $h \circ \eta = \zeta$. Moreover it is evident that h is the only homomorphism of N into K with this property. The proof is therefore complete.

1.5 Symmetric multilinear mappings

Once again M denotes an R-module and all products $M \times M \times \cdots \times M$ and $M \otimes M \otimes \cdots \otimes M$ are understood to have p $(p \geq 1)$ factors.

Let N be an R-module. A multilinear mapping

$$\theta: M \times M \times \cdots \times M \to N$$

is called *symmetric* if

$$\theta(m_1, m_2, \ldots, m_p) = \theta(m_{i_1}, m_{i_2}, \ldots, m_{i_p}), \tag{1.5.1}$$

whenever $m_1, m_2, \ldots, m_p$ belong to M and $(i_1, i_2, \ldots, i_p)$ is a permutation of $(1, 2, \ldots, p)$. Clearly θ is symmetric provided $\theta(m_1, m_2, \ldots, m_p)$ remains unaltered whenever two *adjacent* terms are interchanged.

If θ is a symmetric multilinear mapping and $h: N \to K$ is a homomorphism of R-modules, then $h \circ \theta$ is a symmetric multilinear mapping of $M \times M \times \cdots \times M$ into K. This inevitably prompts the next universal problem.

Problem 3. *To choose N and θ so that given any symmetric multilinear mapping*

$$\omega: M \times M \times \cdots \times M \to K$$

there exists a unique R-homomorphism $h: N \to K$ *such that* $h \circ \theta = \omega$.

As Problem 3 can be treated in very much the same way as Problem 2 there will be no need to go into the same amount of detail.

Denote by $H_p(M)$ the submodule of $T_p(M)$ generated by all differences

$$m_1 \otimes m_2 \otimes \cdots \otimes m_p - m_{i_1} \otimes m_{i_2} \otimes \cdots \otimes m_{i_p}, \tag{1.5.2}$$

where $(i_1, i_2, \ldots, i_p)$ is a permutation of $(1, 2, \ldots, p)$. Put

$$N = T_p(M)/H_p(M)$$

and define $\theta: M \times M \times \cdots \times M \to N$ so that $\theta(m_1, m_2, \ldots, m_p)$ is the natural image of $m_1 \otimes m_2 \otimes \cdots \otimes m_p$ in N. Then θ is multilinear and symmetric. Moreover considerations, closely similar to those encountered in solving Problem 2, show that N and θ solve the present problem. Of course the solution is unique in the same sense and for much the same reasons as apply in the case of the other universal problems.

Now let (N, θ) be *any* solution to Problem 3. We put

$$S_p(M) = N \tag{1.5.3}$$

and

$$m_1 m_2 \ldots m_p = \theta(m_1, m_2, \ldots, m_p). \tag{1.5.4}$$

Then

$$m_1 m_2 \ldots (m_i' + m_i'') \ldots m_p = m_1 m_2 \ldots m_i' \ldots m_p + m_1 m_2 \ldots m_i'' \ldots m_p \tag{1.5.5}$$

and

$$m_1 m_2 \ldots (rm_i) \ldots m_p = r(m_1 m_2 \ldots m_i \ldots m_p) \tag{1.5.6}$$

because θ is multilinear, and

$$m_1 m_2 \ldots m_p = m_{i_1} m_{i_2} \ldots m_{i_p} \tag{1.5.7}$$

because it is symmetric. (Once again $(i_1, i_2, \ldots, i_p)$ denotes an arbitrary permutation of $(1, 2, \ldots, p)$.) The module $S_p(M)$ is called the p-th *symmetric power* of M. As in earlier instances, when $p = 1$ our universal problem is solved by M and its identity mapping. Consequently $S_1(M) = M$. The next theorem records the characteristic property of the general symmetric power.

Theorem 7. *Given an R-module K and a symmetric multilinear mapping*

$$\omega: M \times M \times \cdots \times M \to K$$

there exists a unique R-homomorphism $h: S_p(M) \to K$ *such that*

$$h(m_1 m_2 \ldots m_p) = \omega(m_1, m_2, \ldots, m_p)$$

for all $m_1, m_2, \ldots, m_p$.

Corollary. *Each element of $S_p(M)$ is a finite sum of elements of the form $m_1 m_2 \ldots m_p$.*

As is to be expected a proof of this corollary can be obtained by making minor modifications to the arguments used when establishing the corollary to Theorem 1.

We next observe that, by Theorem 2, there is a homomorphism $T_p(M) \to S_p(M)$, of R-modules, in which $m_1 \otimes m_2 \otimes \cdots \otimes m_p$ is mapped into $m_1 m_2 \ldots m_p$. This is known as the *canonical homomorphism* of $T_p(M)$ into $S_p(M)$. Evidently it is surjective. If we identify $T_1(M)$ and $S_1(M)$ with M, then, for $p=1$, the canonical homomorphism is the identity mapping.

We recall that $H_p(M)$ is the submodule of $T_p(M)$ generated by elements of the form (1.5.2).

Theorem 8. *The canonical homomorphism $T_p(M) \to S_p(M)$ is surjective and its kernel is $H_p(M)$. Furthermore $H_p(M)$ is generated by all differences*

$$m_1 \otimes \cdots \otimes m_i \otimes m_{i+1} \otimes \cdots \otimes m_p$$
$$-m_1 \otimes \cdots \otimes m_{i+1} \otimes m_i \otimes \cdots \otimes m_p,$$

where the second term is obtained from the first by interchanging two adjacent factors.

Here the demonstration parallels the proof of Theorem 5. We leave the reader to make the necessary adjustments.

Finally let us determine the structure of $S_p(M)$ when M is a free R-module with a base B. To do this we consider all sequences $(b_1, b_2, \ldots, b_p)$ of p elements of B (on this occasion repetitions are allowed) and we regard the sequences $(b_1, b_2, \ldots, b_p)$ and $(b'_1, b'_2, \ldots, b'_p)$ as equivalent if each can be obtained by reordering the other. From each equivalence class we now select a representative, and we denote by $I_p^*(B)$ the set consisting of these representatives.

Theorem 9. *Let M be a free R-module with base B. Then $S_p(M)$ is also a free R-module and it has as a base the elements $b_1 b_2 \ldots b_p$, where $(b_1, b_2, \ldots, b_p)$ ranges over $I_p^*(B)$.*

Remark. If the elements of B are regarded as commuting indeterminates, then this result says that $S_p(M)$ can be thought of as consisting of all homogeneous polynomials of degree p, with coefficients in R, in these indeterminates.

Proof. Let N be the free R-module generated by $I_p^*(B)$ and define

$$\theta \colon B \times B \times \cdots \times B \to N$$

by putting $\theta(b_1, b_2, \ldots, b_p) = (b'_1, b'_2, \ldots, b'_p)$, where $(b'_1, b'_2, \ldots, b'_p)$ is the rearrangement of $(b_1, b_2, \ldots, b_p)$ that belongs to $I_p^*(B)$. Naturally θ extends to a multilinear mapping of $M \times M \times \cdots \times M$ into N and it is evident that the multilinear mapping so obtained is symmetric. It now suffices to show that (N, θ) solves Problem 3.

Suppose then that

$$\omega: M \times M \times \cdots \times M \to K$$

is multilinear and symmetric, and define the R-homomorphism $h: N \to K$ so that, for $(b_1, b_2, \ldots, b_p)$ in $I_p^*(B)$, we have $h(b_1, b_2, \ldots, b_p) = \omega(b_1, b_2, \ldots, b_p)$. Then $h \circ \theta$ and ω are symmetric multilinear mappings which agree on $I_p^*(B)$. It follows that they must agree on $B \times B \times \cdots \times B$, and this in turn implies that $h \circ \theta = \omega$. This completes the proof because it is clear that there is only one homomorphism $M \to K$ which, when combined with θ, gives ω.

1.6 Comments and exercises

In this section we shall make some observations to complement what has been said in the main text and we shall provide a number of exercises. A few of these exercises are marked with an asterisk. Those that are so marked are either particularly interesting, or difficult, or used in later chapters. Solutions to the starred questions will be found in Section (1.7).

Throughout Section (1.6) R is assumed to be *non-trivial.*

Since we are concerned with *multilinear algebra* it is fitting that the scope of *linear algebra* (in the present context) should be made explicit. Often by linear algebra is meant the theory of vector spaces and linear transformations. When the term *vector space* is used it is frequently understood that the scalars are taken from a *field.* We, however, will only require our scalars to belong to an arbitrary *commutative ring* and when we make this change the terminology normally changes as well. What had previously been termed a *vector space* now becomes a *module* and what had been described before as a *linear transformation* is now referred to as a *homomorphism.* Thus, for us, *linear algebra* will mean the theory of modules (and their homomorphisms) over a commutative ring.

One of the key problems of linear algebra is to determine when a system of homogeneous linear equations has a non-trivial solution. Let us pose this problem in a very general form.

Suppose that M is an R-module and consider the equations

$$\begin{array}{l} a_{11}m_1 + a_{12}m_2 + \cdots + a_{1q}m_q = 0 \\ a_{21}m_1 + a_{22}m_2 + \cdots + a_{2q}m_q = 0 \\ \quad\vdots \qquad\quad \vdots \qquad\quad \vdots \qquad\quad \vdots \\ a_{p1}m_1 + a_{p2}m_2 + \cdots + a_{pq}m_q = 0. \end{array} \tag{1.6.1}$$

Here the a_{ij} belong to R and we are interested in solving the equations in the module M. Of course one solution is obtained by taking all the m_i to be zero. Any other solution is called *non-trivial.*

The first exercise gives a necessary and sufficient condition for the existence of a non-trivial solution. This result is known as *McCoy's Theorem.* In order to state it we first put

$$A = \begin{bmatrix} a_{11} & a_{12} & \cdots & a_{1q} \\ a_{21} & a_{22} & \cdots & a_{2q} \\ \vdots & \vdots & \vdots\vdots\vdots & \vdots \\ a_{p1} & a_{p2} & \cdots & a_{pq} \end{bmatrix} \tag{1.6.2}$$

so that A is the matrix of coefficients, and we denote by $\mathfrak{A}_\nu(A)$ the ideal that is generated by the $\nu \times \nu$ minors of A. Then

$$\mathfrak{A}_0(A) \supseteq \mathfrak{A}_1(A) \supseteq \mathfrak{A}_2(A) \supseteq \cdots,$$

where by $\mathfrak{A}_0(A)$ we mean the improper ideal R. Note that $\mathfrak{A}_\nu(A) = 0$ for $\nu > \min(p, q)$ because, for such a ν, there are no $\nu \times \nu$ minors.

Exercise 1*. *Show that the equations* (1.6.1) *have a non-trivial solution, in the R-module M, if and only if* $\mathfrak{A}_q(A)$ *annihilates a non-zero element of M. Deduce that if* $M \neq 0$ *and* $p < q$*, then a non-trivial solution exists.*

Exercise 1 can be used *inter alia* to establish some fundamental facts about free modules and their bases. We illustrate this by means of the next exercises.

Exercise 2. *The R-module M can be generated by q elements. Show that any* $q+1$ *elements of M are linearly dependent over R. Show also that if M is a free module with a base of q elements, then* (i) *q is the smallest number of elements which will generate M, and* (ii) *any q elements that generate M form a base.*

Exercise 3. *Let M be a free R-module and let B, B′ be bases of M. Show that either* (i) *B and B′ are both infinite, or* (ii) *B and B′ are both finite and contain the same number of elements.*

Exercise 3 shows that the number of elements in a base of a free R-module M is the same for all choices of the base. This number, which is called the *rank* of the free module, will be denoted by $\operatorname{rank}_R(M)$. Thus $\operatorname{rank}_R(M)$ is either a non-negative integer or it is 'plus infinity'.

We now turn our attention to some other matters.

Exercise 4. *Let* $I_1, I_2, \ldots, I_p$ *be ideals of R such that* $I_1 + I_2 + \cdots + I_p = R$*, and let* $M_j = R/I_j$*. Show that if*

$$\psi: M_1 \times M_2 \times \cdots \times M_p \to N$$

is a multilinear mapping of R-modules, then every element of $M_1 \times M_2 \times \cdots \times M_p$ *is mapped into zero.*

We recall that if M is an R-module and I is an ideal of R, then M/IM has a natural structure as an R/I-module as well as an R-module.

Exercise 5*. *Let* $M_1, M_2, \ldots, M_s$ *be R-modules and let I be an ideal of R. Show that the R/I-modules*

$$(M_1 \otimes_R M_2 \otimes_R \cdots \otimes_R M_s)/I(M_1 \otimes_R M_2 \otimes_R \cdots \otimes_R M_s)$$

and

$$(M_1/IM_1) \otimes_{R/I} (M_2/IM_2) \otimes_{R/I} \cdots \otimes_{R/I} (M_s/IM_s)$$

are isomorphic and exhibit an explicit isomorphism.

In Chapter 5 we shall derive the basic facts about free modules, their bases and their ranks by different means. The central fact in the alternative approach is embodied in the next exercise.

Exercise 6. *Let* $M \neq 0$ *be a free R-module. Show that* $\operatorname{rank}_R(M)$ *is the upper bound of all integers p such that* $E_p(M) \neq 0$.

Exercise 7. *Let M be a free R-module of* rank n. *Determine the ranks of the free modules* $T_p(M)$, $E_p(M)$ *and* $S_p(M)$.

The next exercise shows that it is possible for the tensor square of a non-zero module to be zero.

Exercise 8. *Show that if* $\mathbb{Z}$ *denotes the ring of integers and* $\mathbb{Q}$ *the field of rational numbers, then* $(\mathbb{Q}/\mathbb{Z}) \otimes_{\mathbb{Z}} (\mathbb{Q}/\mathbb{Z}) = 0$.

1.7 Solutions to selected exercises

In this section we shall provide complete solutions to Exercises 1 and 5, and make observations about some of the other exercises.

Exercise 1. *Show that the equations* (1.6.1) *have a non-trivial solution, in the R-module M, if and only if* $\mathfrak{A}_q(M)$ *annihilates a non-zero element of M. Deduce that if* $M \neq 0$ *and* $p < q$, *then a non-trivial solution exists.*

Solution. First suppose that $m_1, m_2, \ldots, m_q$ satisfy the equations with at least one m_i different from zero. *We claim that* $\mathfrak{A}_q(A)m_i = 0$ *for* $i = 1, 2, \ldots, q$. Evidently in establishing this we may suppose that $p \geq q$.

From the first q equations, namely

$$\begin{matrix} a_{11}m_1 + a_{12}m_2 + \cdots + a_{1q}m_q = 0 \\ a_{21}m_1 + a_{22}m_2 + \cdots + a_{2q}m_q = 0 \\ \vdots \\ a_{q1}m_1 + a_{q2}m_2 + \cdots + a_{qq}m_q = 0, \end{matrix}$$

we see that $Dm_i=0$ $(i=1, 2, \ldots, q)$, where D is the $q \times q$ minor

$$\begin{vmatrix} a_{11} & a_{12} & \cdots & a_{1q} \\ a_{21} & a_{22} & \cdots & a_{2q} \\ \vdots & \vdots & \vdots\vdots\vdots & \vdots \\ a_{q1} & a_{q2} & \cdots & a_{qq} \end{vmatrix}.$$

In the same way we can show that any other $q \times q$ minor of A annihilates all the m_i. Accordingly $\mathfrak{A}_q(A)m_i=0$ and the necessity of the stated condition has been established.

Now suppose that $\mathfrak{A}_q(A)$ annihilates a non-zero element of M. If v is the *smallest* integer for which $\mathfrak{A}_v(A)$ annihilates a non-zero element, then $0<v\leq q$ and $\mathfrak{A}_v(A)m=0$ for some $m\neq 0$ in M. If $v=1$ the existence of a non-trivial solution is obvious. Otherwise there is a $(v-1)\times(v-1)$ minor that does not annihilate m. Without loss of generality we can suppose that the $(v-1)\times(v-1)$ minor in question occurs in the top left-hand corner of A.

In the determinant

$$\begin{vmatrix} x_1 & x_2 & \cdots & x_v \\ a_{11} & a_{12} & \cdots & a_{1v} \\ \vdots & \vdots & \vdots\vdots\vdots & \vdots \\ a_{v-1,1} & a_{v-1,2} & \cdots & a_{v-1,v} \end{vmatrix}$$

let c_r be the cofactor of x_r. Then $c_v m\neq 0$. Furthermore, for $1\leq i\leq p$,

$$a_{i1}(c_1m)+a_{i2}(c_2m)+\cdots+a_{iv}(c_vm)$$
$$=m\begin{vmatrix} a_{i1} & a_{i2} & \cdots & a_{iv} \\ a_{11} & a_{12} & \cdots & a_{1v} \\ \vdots & \vdots & \vdots\vdots\vdots & \vdots \\ a_{v-1,1} & a_{v-1,2} & \cdots & a_{v-1,v} \end{vmatrix}.$$

If $1\leq i\leq v-1$ this is zero because the determinant has two equal rows; and if $i\geq v$, then it is zero because $\mathfrak{A}_v(A)m=0$. Thus by taking

$$m_1=c_1m,\ m_2=c_2m, \ldots, m_v=c_vm$$

and

$$m_{v+1}=m_{v+2}=\cdots=m_q=0$$

we obtain a non-trivial solution of the equations.

Finally suppose that $M\neq 0$ and $p<q$. Then $\mathfrak{A}_q(A)=0$ and this certainly annihilates a non-zero element of M. Accordingly a non-trivial solution exists.

Exercise 1 provides the key to Exercise 2.

Exercise 5. *Let $M_1, M_2, \ldots, M_s$ be R-modules and let I be an ideal of R. Show that the R/I-modules*

$$(M_1 \otimes_R M_2 \otimes_R \cdots \otimes_R M_s)/I(M_1 \otimes_R M_2 \otimes_R \cdots \otimes_R M_s)$$

and

$$(M_1/IM_1) \otimes_{R/I} (M_2/IM_2) \otimes_{R/I} \cdots \otimes_{R/I} (M_s/IM_s)$$

are isomorphic and exhibit an explicit isomorphism.

Solution. For $m_j \in M_j$ let $\bar{m}_j$ denote its natural image in M_j/IM_j. Now every R/I-module has an obvious structure as an R-module. On this understanding the mapping

$$M_1 \times M_2 \times \cdots \times M_s \to (M_1/IM_1) \otimes_{R/I} \cdots \otimes_{R/I} (M_s/IM_s)$$

which takes $(m_1, m_2, \ldots, m_s)$ into $\bar{m}_1 \otimes \bar{m}_2 \otimes \cdots \otimes \bar{m}_s$ is a multilinear mapping of R-modules. Accordingly there is induced an R-linear mapping

$$M_1 \otimes_R \cdots \otimes_R M_s \to (M_1/IM_1) \otimes_{R/I} \cdots \otimes_{R/I} (M_s/IM_s)$$

and this, as it clearly vanishes on $I(M_1 \otimes_R \cdots \otimes_R M_s)$, gives rise to a mapping

$$\lambda: (M_1 \otimes_R \cdots \otimes_R M_s)/I(M_1 \otimes_R \cdots \otimes_R M_s)$$
$$\to (M_1/IM_1) \otimes_{R/I} \cdots \otimes_{R/I} (M_s/IM_s).$$

In fact λ is a homomorphism of R/I-modules and if $\overline{m_1 \otimes \cdots \otimes m_s}$ denotes the natural image of $m_1 \otimes \cdots \otimes m_s$ in $(M_1 \otimes_R \cdots \otimes_R M_s)/I(M_1 \otimes_R \cdots \otimes_R M_s)$, then

$$\lambda(\overline{m_1 \otimes \cdots \otimes m_s}) = \bar{m}_1 \otimes \bar{m}_2 \otimes \cdots \otimes \bar{m}_s.$$

Now suppose that, for each j, we have elements m_j and m'_j, in M_j, such that $\bar{m}_j = \bar{m}'_j$. Then

$$m_1 \otimes \cdots \otimes m_j \otimes \cdots \otimes m_s - m_1 \otimes \cdots \otimes m'_j \otimes \cdots \otimes m_s$$
$$= m_1 \otimes \cdots \otimes (m_j - m'_j) \otimes \cdots \otimes m_s$$

and this belongs to $I(M_1 \otimes_R \cdots \otimes_R M_s)$. Repeated applications of this observation show that

$$m_1 \otimes m_2 \otimes \cdots \otimes m_s$$
$$\equiv m'_1 \otimes m'_2 \otimes \cdots \otimes m'_s \pmod{I(M_1 \otimes_R \cdots \otimes_R M_s)}.$$

It follows that there is a well-defined mapping

$$(M_1/IM_1) \times \cdots \times (M_s/IM_s)$$
$$\to (M_1 \otimes_R \cdots \otimes_R M_s)/I(M_1 \otimes_R \cdots \otimes_R M_s)$$

in which $(\bar{m}_1, \bar{m}_2, \ldots, \bar{m}_s)$ goes into $\overline{m_1 \otimes m_2 \otimes \cdots \otimes m_s}$. This is a multilinear mapping of R/I-modules and so it gives rise to an R/I-homomorphism

$$\mu: (M_1/IM_1) \otimes_{R/I} \cdots \otimes_{R/I} (M_s/IM_s)$$
$$\to (M_1 \otimes_R \cdots \otimes_R M_s)/I(M_1 \otimes_R \cdots \otimes_R M_s)$$

which satisfies $\mu(\bar{m}_1 \otimes \cdots \otimes \bar{m}_s) = \overline{(m_1 \otimes m_2 \otimes \cdots \otimes m_s)}$. Finally

$$(\mu \circ \lambda)\overline{(m_1 \otimes \cdots \otimes m_s)} = \mu(\bar{m}_1 \otimes \cdots \otimes \bar{m}_s) = \overline{m_1 \otimes \cdots \otimes m_s}$$

and

$$(\lambda \circ \mu)(\bar{m}_1 \otimes \cdots \otimes \bar{m}_s) = \lambda\overline{(m_1 \otimes \cdots \otimes m_s)} = \bar{m}_1 \otimes \cdots \otimes \bar{m}_s.$$

It follows that λ and μ are inverse isomorphisms and now the solution is complete.

Our concluding observation has to do with the answers to the questions posed by Exercise 7. We record that, for a free R-module M of rank n,

$$\operatorname{rank}_R(T_p(M)) = n^p,$$

$$\operatorname{rank}_R(E_p(M)) = \binom{n}{p},$$

and

$$\operatorname{rank}_R(S_p(M)) = \binom{n-1+p}{p}.$$

2

Some properties of tensor products

General remarks

It was in Section (1.2) that we defined the tensor product of a finite number of modules over a ring R, but at that time we passed on rapidly to study the tensor, exterior and symmetric powers of a module. However, the theory of tensor products is rich in results, some of which will be needed later. Here we provide an account of the most fundamental properties. As in Chapter 1 we shall sometimes omit the subscript to the symbol $\otimes$ if it is clear which is the ring over which the products are formed.

Throughout Chapter 2 the letters R and S denote *commutative* rings each possessing an identity element.

2.1 Basic isomorphisms

Let $M_1, M_2, \ldots, M_p$ and $N_1, N_2, \ldots, N_q$ be R-modules. In what follows $m_1, m_2, \ldots, m_p$ denote elements of $M_1, M_2, \ldots, M_p$ respectively and $n_1, n_2, \ldots, n_q$ denote elements of $N_1, N_2, \ldots, N_q$.

Theorem 1. *There is an R-isomorphism*

$$M_1 \otimes_R \cdots \otimes_R M_p \otimes_R N_1 \otimes_R \cdots \otimes_R N_q \approx (M_1 \otimes_R \cdots \otimes_R M_p) \otimes_R (N_1 \otimes_R \cdots \otimes_R N_q)$$

in which $m_1 \otimes \cdots \otimes m_p \otimes n_1 \otimes \cdots \otimes n_q$ *is matched with* $(m_1 \otimes \cdots \otimes m_p) \otimes (n_1 \otimes \cdots \otimes n_q)$.

Proof. By Theorem 2 of Chapter 1, the multilinear mapping

$$M_1 \times \cdots \times M_p \times N_1 \times \cdots \times N_q \to (M_1 \otimes \cdots \otimes M_p) \otimes (N_1 \otimes \cdots \otimes N_q),$$

where $(m_1, \ldots, m_p, n_1, \ldots, n_q)$ goes to $(m_1 \otimes \cdots \otimes m_p) \otimes (n_1 \otimes \cdots \otimes n_q)$, induces a homomorphism

$$f: M_1 \otimes \cdots \otimes M_p \otimes N_1 \otimes \cdots \otimes N_q \\ \rightarrow (M_1 \otimes \cdots \otimes M_p) \otimes (N_1 \otimes \cdots \otimes N_q)$$

in which

$$f(m_1 \otimes \cdots \otimes m_p \otimes n_1 \otimes \cdots \otimes n_q) \\ = (m_1 \otimes \cdots \otimes m_p) \otimes (n_1 \otimes \cdots \otimes n_q).$$

We now seek to reverse this homomorphism. Suppose for the moment that we keep $n_1, n_2, \ldots, n_q$ *fixed.* Another application of Chapter 1, Theorem 2 yields a homomorphism

$$M_1 \otimes M_2 \otimes \cdots \otimes M_p \rightarrow M_1 \otimes \cdots \otimes M_p \otimes N_1 \otimes \cdots \otimes N_q$$

in which $m_1 \otimes m_2 \otimes \cdots \otimes m_p$ is mapped into $m_1 \otimes \cdots \otimes m_p \otimes n_1 \otimes \cdots \otimes n_q$. Consequently if in $M_1 \otimes M_2 \otimes \cdots \otimes M_p$ we have a relation

$$m_1 \otimes m_2 \otimes \cdots \otimes m_p + m_1' \otimes m_2' \otimes \cdots \otimes m_p' \\ + \cdots + m_1'' \otimes m_2'' \otimes \cdots \otimes m_p'' = 0,$$

then

$$m_1 \otimes \cdots \otimes m_p \otimes n_1 \otimes \cdots \otimes n_q + m_1' \otimes \cdots \otimes m_p' \otimes n_1 \otimes \cdots \otimes n_q \\ + \cdots + m_1'' \otimes \cdots \otimes m_p'' \otimes n_1 \otimes \cdots \otimes n_q$$

is zero in $M_1 \otimes \cdots \otimes M_p \otimes N_1 \otimes \cdots \otimes N_q$. Of course a similar observation applies if the roles of the M_i and N_j are interchanged.

Let $\xi \in M_1 \otimes M_2 \otimes \cdots \otimes M_p$ and $\eta \in N_1 \otimes N_2 \otimes \cdots \otimes N_q$. We know that ξ can be represented as a finite sum of elements of the form $m_1 \otimes m_2 \otimes \cdots \otimes m_p$. Suppose that

$$\xi = \sum m_1 \otimes m_2 \otimes \cdots \otimes m_p = \sum \mu_1 \otimes \mu_2 \otimes \cdots \otimes \mu_p$$

are two such representations of ξ and that

$$\eta = \sum n_1 \otimes n_2 \otimes \cdots \otimes n_q = \sum \nu_1 \otimes \nu_2 \otimes \cdots \otimes \nu_q$$

are two similar representations of η. The remarks in the last paragraph show that

$$\sum\sum m_1 \otimes \cdots \otimes m_p \otimes n_1 \otimes \cdots \otimes n_q \\ = \sum\sum \mu_1 \otimes \cdots \otimes \mu_p \otimes n_1 \otimes \cdots \otimes n_q \\ = \sum\sum \mu_1 \otimes \cdots \otimes \mu_p \otimes \nu_1 \otimes \cdots \otimes \nu_q.$$

Consequently

$$\sum\sum m_1 \otimes \cdots \otimes m_p \otimes n_1 \otimes \cdots \otimes n_q \tag{2.1.1}$$

depends only on ξ and η and not on the chosen representations. It follows that there is a mapping

$$(M_1 \otimes \cdots \otimes M_p) \times (N_1 \otimes \cdots \otimes N_q) \\ \rightarrow M_1 \otimes \cdots \otimes M_p \otimes N_1 \otimes \cdots \otimes N_q$$

that takes (ξ, η) into the element (2.1.1). This mapping, because it is bilinear, induces a homomorphism

$$g: (M_1 \otimes \cdots \otimes M_p) \otimes (N_1 \otimes \cdots \otimes N_q) \\ \rightarrow M_1 \otimes \cdots \otimes M_p \otimes N_1 \otimes \cdots \otimes N_q$$

in which $(m_1 \otimes \cdots \otimes m_p) \otimes (n_1 \otimes \cdots \otimes n_q)$ becomes $m_1 \otimes \cdots \otimes m_p \otimes n_1 \otimes \cdots \otimes n_q$. The theorem follows because $g \circ f$ and $f \circ g$ are clearly identity mappings.

Corollary. *There is an isomorphism*

$$(M_1 \otimes_R M_2) \otimes_R M_3 \approx M_1 \otimes_R (M_2 \otimes_R M_3)$$

of R-modules in which $(m_1 \otimes m_2) \otimes m_3$ *corresponds to* $m_1 \otimes (m_2 \otimes m_3)$.

Proof. The theorem provides isomorphisms

$$(M_1 \otimes M_2) \otimes M_3 \approx M_1 \otimes M_2 \otimes M_3 \approx M_1 \otimes (M_2 \otimes M_3)$$

and the corollary follows by combining them.

Theorem 2. *Let* $(i_1, i_2, \ldots, i_p)$ *be a permutation of* $(1, 2, \ldots, p)$. *Then there is an isomorphism*

$$M_1 \otimes_R M_2 \otimes_R \cdots \otimes_R M_p \approx M_{i_1} \otimes_R M_{i_2} \otimes_R \cdots \otimes_R M_{i_p}$$

of R-modules which associates $m_1 \otimes m_2 \otimes \cdots \otimes m_p$ *with* $m_{i_1} \otimes m_{i_2} \otimes \cdots \otimes m_{i_p}$.

Proof. The multilinear mapping

$$M_1 \times M_2 \times \cdots \times M_p \rightarrow M_{i_1} \otimes M_{i_2} \otimes \cdots \otimes M_{i_p}$$

taking $(m_1, m_2, \ldots, m_p)$ into $m_{i_1} \otimes m_{i_2} \otimes \cdots \otimes m_{i_p}$ induces a homomorphism

$$f: M_1 \otimes M_2 \otimes \cdots \otimes M_p \rightarrow M_{i_1} \otimes M_{i_2} \otimes \cdots \otimes M_{i_p},$$

where $f(m_1 \otimes m_2 \otimes \cdots \otimes m_p) = m_{i_1} \otimes m_{i_2} \otimes \cdots \otimes m_{i_p}$. For similar reasons there is a homomorphism

$$g: M_{i_1} \otimes M_{i_2} \otimes \cdots \otimes M_{i_p} \rightarrow M_1 \otimes M_2 \otimes \cdots \otimes M_p$$

with $g(m_{i_1} \otimes m_{i_2} \otimes \cdots \otimes m_{i_p}) = m_1 \otimes m_2 \otimes \cdots \otimes m_p$. Evidently $g \circ f$ and $f \circ g$ are identity mappings and therefore f is an isomorphism. The theorem is therefore proved.

The ring R is itself an R-module. As the next theorem shows, this particular R-module plays a very special role in the theory of tensor products.

Theorem 3. *Let* M *be an* R*-module. Then there is an isomorphism* $R \otimes_R M \approx M$ *(of R-modules) in which* $r \otimes m$ *is matched with* rm. *There is a similar isomorphism* $M \otimes_R R \approx M$ *matching* $m \otimes r$ *with* rm.

Proof. Theorem 2 of Chapter 1 shows that there is a homomorphism $f: R \otimes_R M \to M$ with $f(r \otimes m) = rm$. The mapping $g: M \to R \otimes_R M$ given by $g(m) = 1 \otimes m$ is certainly a homomorphism. But $r \otimes m = 1 \otimes rm$ and from this it follows that $g \circ f$ and $f \circ g$ are identity mappings. Consequently f is an isomorphism and the theorem is proved.

2.2 Tensor products of homomorphisms

Let $M_1, M_2, \ldots, M_p$ and $M'_1, M'_2, \ldots, M'_p$ be R-modules, and let

$$f_i: M_i \to M'_i \quad (i = 1, 2, \ldots, p) \tag{2.2.1}$$

be given homomorphisms. The mapping

$$M_1 \times M_2 \times \cdots \times M_p \to M'_1 \otimes M'_2 \otimes \cdots \otimes M'_p$$

in which the image of $(m_1, m_2, \ldots, m_p)$ is $f_1(m_1) \otimes f_2(m_2) \otimes \cdots \otimes f_p(m_p)$ is multilinear and therefore it gives rise to a homomorphism

$$M_1 \otimes M_2 \otimes \cdots \otimes M_p \to M'_1 \otimes M'_2 \otimes \cdots \otimes M'_p$$

of R-modules. This homomorphism, which is denoted by $f_1 \otimes f_2 \otimes \cdots \otimes f_p$, satisfies

$$\begin{aligned}(f_1 \otimes f_2 \otimes \cdots \otimes f_p)&(m_1 \otimes m_2 \otimes \cdots \otimes m_p) \\ &= f_1(m_1) \otimes f_2(m_2) \otimes \cdots \otimes f_p(m_p).\end{aligned} \tag{2.2.2}$$

Note that if each f_i is surjective, then $f_1 \otimes f_2 \otimes \cdots \otimes f_p$ is surjective as well. Note too that if $M'_i = M_i$ for $i = 1, 2, \ldots, p$ and f_i is the identity mapping of M_i, then $f_1 \otimes f_2 \otimes \cdots \otimes f_p$ is the identity mapping of $M_1 \otimes M_2 \otimes \cdots \otimes M_p$.

Once again suppose that homomorphisms (2.2.1) are given and that, in addition to these, we have further homomorphisms

$$g_i: M''_i \to M_i \quad (i = 1, 2, \ldots, p). \tag{2.2.3}$$

It then follows, from (2.2.2), that

$$\begin{aligned}(f_1 \otimes f_2 \otimes \cdots \otimes f_p)&\circ(g_1 \otimes g_2 \otimes \cdots \otimes g_p) \\ &= (f_1 \circ g_1) \otimes (f_2 \circ g_2) \otimes \cdots \otimes (f_p \circ g_p).\end{aligned} \tag{2.2.4}$$

We can deduce from this that when each f_i is an isomorphism, then $f_1 \otimes f_2 \otimes \cdots \otimes f_p$ is also an isomorphism. For in this situation we can form both $f_1 \otimes f_2 \otimes \cdots \otimes f_p$ and $f_1^{-1} \otimes f_2^{-1} \otimes \cdots \otimes f_p^{-1}$, and when these are combined, in either order, (2.2.4) shows that we obtain an identity mapping. Consequently not only is $f_1 \otimes f_2 \otimes \cdots \otimes f_p$ an isomorphism, but we also have

$$(f_1 \otimes f_2 \otimes \cdots \otimes f_p)^{-1} = f_1^{-1} \otimes f_2^{-1} \otimes \cdots \otimes f_p^{-1}. \tag{2.2.5}$$

(The reader who is acquainted with the language of Category Theory will recognize that much of what has so far been said in this section can be summarized succinctly by the statement that the p-fold tensor product is a *covariant functor* of p variables.)

There is an additional item of notation which it is convenient to record here. Suppose that $f_1, \ldots, f_{i-1}, f_{i+1}, \ldots, f_p$ are the identity mappings of $M_1, \ldots, M_{i-1}, M_{i+1}, \ldots, M_p$ respectively, and that $f_i: M_i \to M_i'$ is a general homomorphism. Under these conditions $f_1 \otimes f_2 \otimes \cdots \otimes f_p$ is often written as $M_1 \otimes \cdots \otimes f_i \otimes \cdots \otimes M_p$. Thus

$$(M_1 \otimes \cdots \otimes f_i \otimes \cdots \otimes M_p)(m_1 \otimes \cdots \otimes m_i \otimes \cdots \otimes m_p) \\ = m_1 \otimes \cdots \otimes f_i(m_i) \otimes \cdots \otimes m_p. \tag{2.2.6}$$

We recall that if f, g are homomorphisms of an R-module M into an R-module M', and if $r \in R$, then we can form new homomorphisms $f+g$ and rf of M into M'. Indeed to be explicit $(f+g)(m) = f(m) + g(m)$ and $(rf)(m) = rf(m)$. In this connection we note that

$$f_1 \otimes \cdots \otimes (f_i' + f_i'') \otimes \cdots \otimes f_p \\ = f_1 \otimes \cdots \otimes f_i' \otimes \cdots \otimes f_p + f_1 \otimes \cdots \otimes f_i'' \otimes \cdots \otimes f_p \tag{2.2.7}$$

and

$$f_1 \otimes \cdots \otimes rf_i \otimes \cdots \otimes f_p = r(f_1 \otimes \cdots \otimes f_i \otimes \cdots \otimes f_p), \tag{2.2.8}$$

where the notation is self-explanatory.

After these preliminaries we turn our attention to the behaviour of tensor products in relation to exact sequences of the form

$$M_i'' \xrightarrow{g_i} M_i \xrightarrow{f_i} M_i' \longrightarrow 0 \tag{2.2.9}$$

where $i = 1, 2, \ldots, p$. Denote by N_i the image of $M_1 \otimes \cdots \otimes g_i \otimes \cdots \otimes M_p$ in $M_1 \otimes M_2 \otimes \cdots \otimes M_p$. Thus N_i is generated by the elements of the form $m_1 \otimes \cdots \otimes g_i(m_i'') \otimes \cdots \otimes m_p$ and, in particular, it is contained in the kernel of $f_1 \otimes f_2 \otimes \cdots \otimes f_p$.

Theorem 4. *Suppose that for each i $(1 \leq i \leq p)$ the sequence* (2.2.9) *is exact. Then the homomorphism*

$$f_1 \otimes f_2 \otimes \cdots \otimes f_p: M_1 \otimes_R M_2 \otimes_R \cdots \otimes_R M_p \\ \to M_1' \otimes_R M_2' \otimes_R \cdots \otimes_R M_p'$$

is surjective and its kernel is $N_1 + N_2 + \cdots + N_p$, where N_i is the image of $M_1 \otimes \cdots \otimes g_i \otimes \cdots \otimes M_p$ in $M_1 \otimes_R M_2 \otimes_R \cdots \otimes_R M_p$.

Proof. For $1 \leq i \leq p$ let $m_i' \in M_i'$ and suppose that m_i and μ_i belong to M_i and are such that $f_i(m_i) = f_i(\mu_i) = m_i'$. Then there exists $m_i'' \in M_i''$ such that $m_i - \mu_i = g_i(m_i'')$ and so we see that

$$m_1 \otimes \cdots \otimes m_i \otimes \cdots \otimes m_p - m_1 \otimes \cdots \otimes \mu_i \otimes \cdots \otimes m_p$$

is in N_i. It follows that

$$m_1 \otimes m_2 \otimes \cdots \otimes m_p - \mu_1 \otimes \mu_2 \otimes \cdots \otimes \mu_p$$

is in $N_1 + N_2 + \cdots + N_p$ and therefore the natural image of $m_1 \otimes m_2 \otimes \cdots \otimes m_p$ in

$$(M_1 \otimes M_2 \otimes \cdots \otimes M_p)/(N_1 + N_2 + \cdots + N_p)$$

depends only on $(m'_1, m'_2, \ldots, m'_p)$. It is now easily checked that there is a multilinear mapping

$$M'_1 \times M'_2 \times \cdots \times M'_p \\ \to (M_1 \otimes M_2 \otimes \cdots \otimes M_p)/(N_1 + N_2 + \cdots + N_p)$$

which is such that $(m'_1, m'_2, \ldots, m'_p)$ is mapped into the image in question. In this way we arrive at a homomorphism

$$M'_1 \otimes M'_2 \otimes \cdots \otimes M'_p \\ \to (M_1 \otimes M_2 \otimes \cdots \otimes M_p)/(N_1 + N_2 + \cdots + N_p) \qquad (2.2.10)$$

that takes $m'_1 \otimes m'_2 \otimes \cdots \otimes m'_p$ into the natural image of $m_1 \otimes m_2 \otimes \cdots \otimes m_p$ in the factor module.

Consider the homomorphisms

$$M_1 \otimes \cdots \otimes M_p \to M'_1 \otimes \cdots \otimes M'_p \\ \to (M_1 \otimes \cdots \otimes M_p)/(N_1 + \cdots + N_p)$$

where the first is $f_1 \otimes f_2 \otimes \cdots \otimes f_p$ and the second is (2.2.10). Their combined effect is to reproduce the natural mapping of $M_1 \otimes M_2 \otimes \cdots \otimes M_p$ onto $(M_1 \otimes M_2 \otimes \cdots \otimes M_p)/(N_1 + N_2 + \cdots + N_p)$ so the kernel of $f_1 \otimes f_2 \otimes \cdots \otimes f_p$ must be contained in $N_1 + N_2 + \cdots + N_p$. But we also have the opposite inclusion because, as was noted earlier, each N_i is contained in the kernel. The proof is now complete.

Corollary. *Let $M''_\mu \to M_\mu \to M'_\mu \to 0$ be an exact sequence of R-modules. Then the induced sequence*

$$M_1 \otimes_R \cdots \otimes_R M''_\mu \otimes_R \cdots \otimes_R M_p \\ \to M_1 \otimes_R \cdots \otimes_R M_\mu \otimes_R \cdots \otimes_R M_p \\ \to M_1 \otimes_R \cdots \otimes_R M'_\mu \otimes_R \cdots \otimes_R M_p \to 0$$

is also exact.

Proof. The corollary is derived from the theorem by taking all the f_i, with the exception of f_μ, to be identity mappings. This ensures that $N_i = 0$ for all $i \neq \mu$.

The property described in the corollary is usually referred to by saying that tensor products are *right exact.*

2.3 Tensor products and direct sums

Let $\{N_i\}_{i \in I}$ be a family of submodules of an R-module N. Then N is the *direct sum* of these submodules if each $n \in N$ has a unique representation of the form

$$n = \sum_{i \in I} n_i, \tag{2.3.1}$$

where $n_i \in N_i$ and only finitely many summands are non-zero. When this is the case we shall usually write

$$N = \sum_{i \in I} N_i \quad \text{(d.s.)}. \tag{2.3.2}$$

However, when N is the direct sum of a finite number of submodules, say of $N_1, N_2, \ldots, N_s$, we shall use

$$N = N_1 \oplus N_2 \oplus \cdots \oplus N_s \tag{2.3.3}$$

as an alternative.

Suppose that we have the situation envisaged in (2.3.2). Then for each $i \in I$ we have an *inclusion mapping* $\sigma_i \colon N_i \to N$ and a *projection mapping* $\pi_i \colon N \to N_i$. (For $n \in N$ the latter picks out from the representation (2.3.1) the summand indexed by i.) Both σ_i and π_i are R-homomorphisms. They have the properties listed below:

(A) *$\pi_j \circ \sigma_i$ is a null homomorphism if $i \neq j$ and it is the identity mapping of N_i if $i = j$;*

(B) *for each $n \in N$, $\pi_i(n)$ is non-zero for only finitely many different values of i;*

(C) *for each $n \in N$, $\sum_{i \in I} \sigma_i \pi_i(n) = n$.*

Let us now make a completely fresh start. Suppose that N is an R-module, and that $\{N_i\}_{i \in I}$ is a family of R-modules which, however, are no longer assumed to be submodules of N. Suppose too that, for each $i \in I$, we are given homomorphisms $\sigma_i \colon N_i \to N$, $\pi_i \colon N \to N_i$ and that these satisfy conditions (A), (B) and (C).

Since $\pi_i \circ \sigma_i$ is the identity mapping of N_i, σ_i is an injection and π_i is a surjection. In particular σ_i maps N_i isomorphically onto $\sigma_i(N_i)$. Next from (B) and (C) it follows that

$$N = \sum_{i \in I} \sigma_i(N_i)$$

and now, by using (A), we can deduce that N is the direct sum of its submodules $\{\sigma_i(N_i)\}_{i \in I}$ in the sense used at the beginning of this section. Thus we have slightly generalized the notion of a direct sum. We shall call the system formed by N, the N_i, and the homomorphisms σ_i and π_i, a *complete representation of* N *as a direct sum.* Furthermore we shall continue to write

$$N=\sum_{i\in I} N_i \quad \text{(d.s.)}$$

in the general case, and $N=N_1\oplus N_2\oplus\cdots\oplus N_s$ in the case where the family $\{N_i\}_{i\in I}$ reduces to $\{N_1, N_2, \ldots, N_s\}$.

Now suppose that $M_1, M_2, \ldots, M_p$ are R-modules and that each is given a (generalized) direct sum decomposition, say

$$M_\mu=\sum_{i_\mu\in I_\mu} M_{i_\mu} \quad \text{(d.s.)},$$

where the complete representation is provided by homomorphisms

$$\sigma_{i_\mu}: M_{i_\mu}\to M_\mu \quad (i_\mu\in I_\mu)$$

and

$$\pi_{i_\mu}: M_\mu\to M_{i_\mu} \quad (i_\mu\in I_\mu).$$

Theorem 5. *Suppose that for each* μ $(1\leqslant\mu\leqslant p)$ *we have*

$$M_\mu=\sum_{i_\mu\in I_\mu} M_{i_\mu} \quad \text{(d.s.)},$$

where the details of the complete representation are as described above. Then

$$M_1\otimes_R M_2\otimes_R\cdots\otimes_R M_p$$

$$=\sum_{(i_1,\ldots,i_p)} M_{i_1}\otimes_R M_{i_2}\otimes_R\cdots\otimes_R M_{i_p} \quad \text{(d.s.)},$$

where the typical homomorphisms in the complete representation are $\sigma_{i_1}\otimes\sigma_{i_2}\otimes\cdots\otimes\sigma_{i_p}$ *and* $\pi_{i_1}\otimes\pi_{i_2}\otimes\cdots\otimes\pi_{i_p}$.

Proof. Put $I=I_1\times I_2\times\cdots\times I_p$, $N=M_1\otimes M_2\otimes\cdots\otimes M_p$ and, for $i=(i_1, i_2, \ldots, i_p)$ in I, set

$$N_i=M_{i_1}\otimes M_{i_2}\otimes\cdots\otimes M_{i_p},$$

$\sigma_i=\sigma_{i_1}\otimes\sigma_{i_2}\otimes\cdots\otimes\sigma_{i_p}$ and $\pi_i=\pi_{i_1}\otimes\pi_{i_2}\otimes\cdots\otimes\pi_{i_p}$. Then $\sigma_i: N_i\to N$, $\pi_i: N\to N_i$ and to prove the theorem it will suffice to show that the conditions which were listed earlier as (A), (B), (C) all hold. That (A) holds is an immediate consequence of (2.2.4). Indeed (B) and (C) are equally obvious as soon as it is realized that we need only verify them when n has the form $m_1\otimes m_2\otimes\cdots\otimes m_p$.

Corollary. *Let* F *be a free* R*-module with a base* $\{\xi_i\}_{i\in I}$ *and let* N *be an arbitrary* R*-module. Then each element of* $F\otimes_R N$ *has a unique representation in the form* $\sum(\xi_i\otimes n_i)$, *where the* n_i *belong to* N *and only finitely many of them are non-zero.*

Proof. We have

$$F = \sum_{i \in I} R\xi_i \quad \text{(d.s.)}$$

and N is also a direct sum with N itself as the only summand. Accordingly, by the theorem,

$$F \otimes N = \sum_{i \in I} (R\xi_i \otimes N) \quad \text{(d.s.)}$$

and so, to establish the corollary, we need only show that for each x in $R\xi_i \otimes N$ there is a unique $n \in N$ such that $x = \xi_i \otimes n$.

By Theorem 3, there is an isomorphism $N \approx R \otimes N$ in which n corresponds to $1 \otimes n$, and of course there is an isomorphism $R \otimes N \approx R\xi_i \otimes N$ which matches $1 \otimes n$ with $\xi_i \otimes n$. Together these provide an isomorphism $N \approx R\xi_i \otimes N$ with n corresponding to $\xi_i \otimes n$. Consequently, as n ranges over N, $\xi_i \otimes n$ ranges over $R\xi_i \otimes N$ giving each element once and once only.

We recall that a module which is a direct summand of a free module is called a *projective* module.

Theorem 6. *Let $0 \to M'' \to M \to M' \to 0$ be an exact sequence of R-modules and let P be a projective R-module. Then the induced sequences*

$$0 \to P \otimes_R M'' \to P \otimes_R M \to P \otimes_R M' \to 0$$

and

$$0 \to M'' \otimes_R P \to M \otimes_R P \to M' \otimes_R P \to 0$$

are exact.

Proof. Of course we need only consider the first sequence. Let $f: M'' \to M$ be an injective homomorphism. By Theorem 4 Cor., it will suffice to show that $P \otimes f$ is also injective. We begin with the case where P is free.

Suppose that F is a free R-module and that $\{\xi_i\}_{i \in I}$ is one of its bases. Let $x \in F \otimes M''$. By Theorem 5 Cor., $x = \sum (\xi_i \otimes m_i'')$, where the m_i'' are in M'' and only finitely many of them are non-zero. Assume that $(F \otimes f)(x) = 0$. Then

$$\sum (\xi_i \otimes f(m_i'')) = 0$$

and therefore $f(m_i'') = 0$ for all $i \in I$, again by the corollary to Theorem 5. But f is an injection. Consequently all the m_i'' are zero and hence $x = 0$. This proves that $F \otimes f$ is an injection.

Finally assume that $F = P \oplus Q$, where F is free and P and Q are submodules. There exist homomorphisms $\sigma: P \to F$ and $\pi: F \to P$ such that $\pi \circ \sigma$ is the identity mapping of P. Now $(\pi \otimes M'') \circ (\sigma \otimes M'') = (\pi \circ \sigma) \otimes M''$ and this is the identity mapping of $P \otimes M''$. It follows that $\sigma \otimes M''$ is an injection. But we already know that $F \otimes f$ is an injection so $(F\otimes f)\circ(\sigma\otimes M'')=\sigma\otimes f$ is an injection as well. Since $\sigma\otimes f=(\sigma \otimes M)\circ(P \otimes f)$, this implies that $P \otimes f$ is an injection, which is what we were aiming to prove.

2.4 Additional structure

In Section (2.4) S, as well as R, denotes a commutative ring with an identity element. If M is both an R-module and an S-module, then we shall say it is an *(R, S)-module* provided (i) the sum of two elements of M is the same in both structures, and (ii) $s(rm)=r(sm)$ whenever $r\in R$, $s\in S$ and $m\in M$. Let M and M' be (R, S)-modules. A mapping $f: M\rightarrow M'$ is called an *(R, S)-homomorphism* if it is a homomorphism of R-modules and also a homomorphism of S-modules. A bijective (R, S)-homomorphism is, of course, referred to as an *(R, S)-isomorphism.*

Suppose that $M_1,\ldots, M_{\mu-1}, M_{\mu+1},\ldots, M_p$ are R-modules and that M_μ is an (R, S)-module. For $s\in S$ define f_s: $M_\mu\rightarrow M_\mu$ by $f_s(m_\mu)=sm_\mu$. Then f_s is an R-homomorphism and therefore $M_1\otimes\cdots\otimes f_s\otimes\cdots\otimes M_p$, where f_s occurs in the μ-th position, is an endomorphism of the R-module $M_1\otimes\cdots\otimes M_\mu\otimes\cdots\otimes M_p$. For x in $M_1\otimes\cdots\otimes M_\mu\otimes\cdots\otimes M_p$ define sx by

$$sx=(M_1\otimes\cdots\otimes f_s\otimes\cdots\otimes M_p)(x). \tag{2.4.1}$$

Then, because $f_{s+\sigma}=f_s+f_\sigma$ and $f_{s\sigma}=f_s\circ f_\sigma$ for $s, \sigma\in S$, it is easily verified that $M_1\otimes\cdots\otimes M_\mu\otimes\cdots\otimes M_p$ is an S-module. Indeed we can go further and say that it is actually an (R, S)-module with

$$s(m_1\otimes\cdots\otimes m_\mu\otimes\cdots\otimes m_p)=m_1\otimes\cdots\otimes sm_\mu\otimes\cdots\otimes m_p. \tag{2.4.2}$$

Finally, if g_i: $M_i\rightarrow M_i'$ is an R-homomorphism for $i=1,\ldots,\mu-1, \mu+1, \ldots, p$ and g_μ: $M_\mu\rightarrow M_\mu'$ is an (R, S)-homomorphism, then $g_1\otimes\cdots\otimes g_\mu\otimes\cdots\otimes g_p$ is itself an (R, S)-homomorphism.

After these preliminaries let A be an R-module, C an S-module, and B an (R, S)-module. The considerations set out in the earlier paragraphs show that $A \otimes_R B$ is an (R, S)-module and therefore $(A \otimes_R B)\otimes_S C$ is an (R, S)-module. Similar considerations show that $A\otimes_R(B\otimes_S C)$ is an (R, S)-module as well. The next theorem, which asserts that $(A\otimes_R B)\otimes_S C$ and $A\otimes_R(B\otimes_S C)$ are virtually identical bimodules, is known as the *associative law* for tensor products.

Theorem 7. *Let A be an R-module, B an (R, S)-module, and C an S-module. Then there is an (R, S)-isomorphism*

$$(A\otimes_R B)\otimes_S C\approx A\otimes_R(B\otimes_S C)$$

which matches $(a\otimes b)\otimes c$ with $a\otimes(b\otimes c)$.

Proof. We borrow an idea from the proof of Theorem 1. Select an element c from C and for the moment let it be kept *fixed*. Consider the mapping

$$A \times B \to A \otimes_R (B \otimes_S C),$$

where (a, b) is mapped into $a \otimes (b \otimes c)$. If we regard A, B and $A \otimes_R (B \otimes_S C)$ simply as R-modules, then the mapping is bilinear and therefore it induces a homomorphism

$$A \otimes_R B \to A \otimes_R (B \otimes_S C)$$

in which $a \otimes b$ goes into $a \otimes (b \otimes c)$. It follows that if

$$a \otimes b + a' \otimes b' + a'' \otimes b'' + \cdots = 0$$

in $A \otimes_R B$, then

$$a \otimes (b \otimes c) + a' \otimes (b' \otimes c) + a'' \otimes (b'' \otimes c) + \cdots = 0$$

in $A \otimes_R (B \otimes_S C)$.

Let x belong to $A \otimes_R B$. Then x can be represented in the form

$$x = a_1 \otimes b_1 + a_2 \otimes b_2 + \cdots + a_n \otimes b_n$$

and the remarks of the last paragraph show that

$$a_1 \otimes (b_1 \otimes c) + a_2 \otimes (b_2 \otimes c) + \cdots + a_n \otimes (b_n \otimes c) \tag{2.4.3}$$

depends only on x and c and is independent of the choice of the representation of x. Consequently there is a well-defined mapping of $(A \otimes_R B) \times C$ into $A \otimes_R (B \otimes_S C)$ in which (x, c) is mapped into the element (2.4.3). But if we regard $A \otimes_R B$, C and $A \otimes_R (B \otimes_S C)$ as S-modules, then this is a bilinear mapping and therefore it induces an S-homomorphism

$$\lambda: (A \otimes_R B) \otimes_S C \to A \otimes_R (B \otimes_S C)$$

with

$$\lambda((a \otimes b) \otimes c) = a \otimes (b \otimes c). \tag{2.4.4}$$

It is now clear, from (2.4.4), that λ is actually an (R, S)-homomorphism.

Entirely similar considerations show that there is an (R, S)-homomorphism

$$\mu: A \otimes_R (B \otimes_S C) \to (A \otimes_R B) \otimes_S C$$

for which $\mu(a \otimes (b \otimes c)) = (a \otimes b) \otimes c$. Since $\mu \circ \lambda$ and $\lambda \circ \mu$ are identity mappings, λ and μ are inverse isomorphisms. The proof is therefore complete.

2.5 Covariant extension

Once again R and S denote commutative rings, only now we assume that a ring-homomorphism

$$\omega: R \to S \tag{2.5.1}$$

(taking identity element into identity element) has been given. Any S-

module U can be regarded as an R-module by defining ru, where $r \in R$ and $u \in U$, to be the same as $\omega(r)u$. In fact this turns U into an (R, S)-module. It follows, as was shown in the last section, that if M is an R-module, then $M \otimes_R U$ is an (R, S)-module. Note that it is also true of $M \otimes_R U$ that multiplication of an element by r is the same as multiplication by $\omega(r)$.

A special case arises if we take U to be S itself. Thus if M is an R-module, then $M \otimes_R S$ is an (R, S)-module. Actually it is the S-module structure that concerns us more because R operates through S in the simple manner just explained. When $M \otimes_R S$ is considered as an S-module we call it the *covariant extension* of M by means of the ring-homomorphism (2.5.1). Covariant extensions are important because they cover such basic constructions as *localization* and the formation of *fractions* and *polynomials*. By combining these constructions under a general heading we can derive some of their most important properties while still keeping the details comparatively simple.

Theorem 8. *Let M be a free R-module with $\{\xi_i\}_{i \in I}$ as a base. If now the ring S is non-trivial, then the covariant extension $M \otimes_R S$ is a free S-module and it has $\{\xi_i \otimes 1\}_{i \in I}$ as a base.*

Proof. Let $x \in M \otimes_R S$. By Theorem 5 Cor., there is a unique family $\{s_i\}_{i \in I}$, of elements of S, with only finitely many s_i non-zero and $x = \sum (\xi_i \otimes s_i)$. The theorem follows because $\xi_i \otimes s_i = s_i(\xi_i \otimes 1)$.

Corollary. *The covariant extension of a projective R-module is a projective S-module.*

Proof. Suppose that F is a free R-module and that P, Q are R-submodules with $F = P \oplus Q$. By Theorem 5,

$$F \otimes_R S = (P \otimes_R S) \oplus (Q \otimes_R S), \tag{2.5.2}$$

this being, in the first instance, a direct sum of R-modules. However, $F \otimes_R S$, $P \otimes_R S$ and $Q \otimes_R S$ are all of them S-modules, and now (2.5.2) is seen to be a direct sum of S-modules. But it has just been shown that $F \otimes_R S$ is S-free. It therefore follows that $P \otimes_R S$ and $Q \otimes_R S$ are S-projective.

Our final result in this chapter shows that tensor products and covariant extensions commute.

Theorem 9. *Let $M_1, M_2, \ldots, M_p$ be R-modules. Then there is an isomorphism*

$$\begin{aligned}(M_1 \otimes_R S) \otimes_S (M_2 \otimes_R S) \otimes_S \cdots \otimes_S (M_p \otimes_R S)\\ \approx (M_1 \otimes_R M_2 \otimes_R \cdots \otimes_R M_p) \otimes_R S\end{aligned}$$

of S-modules in which $(m_1 \otimes s_1) \otimes (m_2 \otimes s_2) \otimes \cdots \otimes (m_p \otimes s_p)$ corresponds to $(m_1 \otimes m_2 \otimes \cdots \otimes m_p) \otimes s_1 s_2 \ldots s_p$.

Proof. We begin with the case $p=2$. By Theorem 7, there is an (R, S)-isomorphism

$$(M_1 \otimes_R S) \otimes_S (M_2 \otimes_R S) \approx M_1 \otimes_R (S \otimes_S (M_2 \otimes_R S)) \tag{2.5.3}$$

which matches $(m_1 \otimes s_1) \otimes (m_2 \otimes s_2)$ with $m_1 \otimes (s_1 \otimes (m_2 \otimes s_2))$. Next, Theorem 3 provides an S-isomorphism

$$S \otimes_S (M_2 \otimes_R S) \approx M_2 \otimes_R S \tag{2.5.4}$$

which pairs $s_1 \otimes (m_2 \otimes s_2)$ with $m_2 \otimes s_1 s_2$. Indeed this is also an (R, S)-isomorphism. Thus (2.5.3) and (2.5.4) together provide an (R, S)-isomorphism

$$(M_1 \otimes_R S) \otimes_S (M_2 \otimes_R S) \approx M_1 \otimes_R (M_2 \otimes_R S)$$

in which $(m_1 \otimes s_1) \otimes (m_2 \otimes s_2)$ goes into $m_1 \otimes (m_2 \otimes s_1 s_2)$. Finally, the corollary to Theorem 1 provides an R-isomorphism

$$M_1 \otimes_R (M_2 \otimes_R S) \approx (M_1 \otimes_R M_2) \otimes_R S, \tag{2.5.5}$$

where $m_1 \otimes (m_2 \otimes s)$ is associated with $(m_1 \otimes m_2) \otimes s$. The case $p=2$ follows as soon as it is noted that (2.5.5) is an S-isomorphism as well as an R-isomorphism.

When $p=1$ the theorem becomes a tautology. The general result follows by induction if we make use of Theorem 1 and the case where $p=2$.

2.6 Comments and exercises

This section will be used to clarify certain aspects of the theory of tensor products by means of comments and exercises. As in the last chapter solutions are provided to the more interesting and difficult exercises, and the exercises which come into this category are marked by an asterisk.

Let M' be a submodule of an R-module M and N' a submodule of an R-module N. The purpose of the first exercise is to show that usually $M' \otimes N'$ cannot be regarded as a submodule of $M \otimes N$. More precisely, the inclusion mappings $M' \to M$ and $N' \to N$ induce a homomorphism

$$M' \otimes N' \to M \otimes N, \tag{2.6.1}$$

and now the point being made is covered by

Exercise 1. *Show by means of an example that the homomorphism $M' \otimes N' \to M \otimes N$ of* (2.6.1) *need not be an injection.*
(Hint. Take $R = \mathbb{Z}$, $M = M' = \mathbb{Z}/2\mathbb{Z}$, $N = \mathbb{Z}$, and $N' = 2\mathbb{Z}$. As always $\mathbb{Z}$ denotes the ring of integers.)

By contrast Exercise 2 embodies an important constructive result which partially offsets the purely negative assertion of Exercise 1.

Exercise 2*. *Let $x_1, x_2, \ldots, x_s$ be elements of an R-module M, and let*

$y_1, y_2, \ldots, y_s$ belong to an R-module N. Suppose that $x_1 \otimes y_1 + x_2 \otimes y_2 + \cdots + x_s \otimes y_s = 0$ in $M \otimes N$. Show that there exist finitely generated submodules M' and N', of M and N respectively, such that

(i) *the x_i are in M' and the y_i in N',*

and

(ii) *$x_1 \otimes y_1 + x_2 \otimes y_2 + \cdots + x_s \otimes y_s = 0$ in $M' \otimes N'$.*

Exercise 3*. *Let I be an ideal of R and let M be an R-module. Establish an isomorphism $M \otimes_R (R/I) \approx M/IM$ of R-modules.*

The next exercise provides a version of a very useful result known as *Nakayama's Lemma*. It appears here because it helps to solve Exercise 5.

Exercise 4*. *Let M be a finitely generated R-module. Show that if $M = JM$ for every maximal ideal J, then $M = 0$.*

The condition that M be finitely generated cannot be dropped. This may be seen by taking R to be $\mathbb{Z}$ and letting M be the field of rational numbers considered as a $\mathbb{Z}$-module.

Exercise 5*. *Let M be an R-module. Show that the following two statements are equivalent:*

(i) *for a finitely generated R-module N we have $M \otimes N = 0$ only when $N = 0$;*

(ii) *for every maximal ideal J, of R, JM is different from M.*

We next consider a matter that concerns tensor products of homomorphisms. To this end let $f: M \to N$ and $f': M' \to N'$ be homomorphisms of R-modules. Then, as we saw in Section (2.2), these induce a homomorphism

$$f \otimes f': M \otimes M' \to N \otimes N'.$$

Now the homomorphisms of M into N can be added and they can be multiplied by elements of R to produce new homomorphisms; in fact the homomorphisms of M into N form an R-module. This module is denoted by $\mathrm{Hom}_R(M, N)$ and this is sometimes simplified to $\mathrm{Hom}(M, N)$ if there is no uncertainty as to the ring of scalars. Since $f \otimes f'$ belongs to $\mathrm{Hom}(M \otimes M', N \otimes N')$, we have a mapping

$$\mathrm{Hom}(M, N) \times \mathrm{Hom}(M', N') \to \mathrm{Hom}(M \otimes M', N \otimes N')$$

in which (f, f') is taken into $f \otimes f'$, and this mapping is bilinear (see (2.2.7) and (2.2.8)). It therefore induces a homomorphism

$$\mathrm{Hom}(M, N) \otimes \mathrm{Hom}(M', N') \to \mathrm{Hom}(M \otimes M', N \otimes N') \qquad (2.6.2)$$

of R-modules.

At this point we encounter an ambiguity in our notation because $f \otimes f'$

could denote either an element of $\text{Hom}(M, N) \otimes \text{Hom}(M', N')$ or an element of $\text{Hom}(M \otimes M', N \otimes N')$; indeed what the mapping (2.6.2) accomplishes is to change the former interpretation into the latter. In practice this double meaning does not cause problems. Usually $f \otimes f'$ means the homomorphism of $M \otimes M'$ into $N \otimes N'$ and in any event the element of doubt can be removed by examining the context. The next exercise has been included to emphasize the difference between the two meanings of $f \otimes f'$.

Exercise 6. *Let R be the ring $\mathbb{Z}/4\mathbb{Z}$ and let $I = 2\mathbb{Z}/4\mathbb{Z}$ so that I is an ideal of R. Put $M = M' = R/I$ and $N = N' = R$. Show that in this case the homomorphism*

$$\text{Hom}(M, N) \otimes \text{Hom}(M', N') \to \text{Hom}(M \otimes M', N \otimes N')$$

of (2.6.2) *is neither an injection nor a surjection.*

We know, from Theorem 4 Cor., that tensor products are right exact. Thus if $0 \to E' \to E \to E'' \to 0$ is an exact sequence of R-modules and M is an arbitrary R-module, then the derived sequence

$$M \otimes E' \to M \otimes E \to M \otimes E'' \to 0$$

is exact. However, for some choices of M it can happen that the five-term sequence

$$0 \to M \otimes E' \to M \otimes E \to M \otimes E'' \to 0$$

is not exact. This observation leads on to the definition of an importance class of R-modules.

Definition. *The R-module M is said to be 'flat' or 'R-flat' if whenever $0 \to E' \to E \to E'' \to 0$ is an exact sequence of R-modules the resulting sequences*

$$0 \to M \otimes E' \to M \otimes E \to M \otimes E'' \to 0 \tag{2.6.3}$$

and

$$0 \to E' \otimes M \to E \otimes M \to E'' \otimes M \to 0 \tag{2.6.4}$$

are exact.

The reader should note that *two* sequences have been mentioned in the interests of symmetry. It is very easy to show, using Theorem 2, that if either one of (2.6.3) and (2.6.4) is exact, then so too is the other. An alternative (but equivalent) way of defining flat R-modules is the following: *M is flat if and only if $M \otimes f$ and $f \otimes M$ are injections whenever f is an injective homomorphism.* Of course, if either $f \otimes M$ or $M \otimes f$ is an injection, then the other is an injection as well. This again is a consequence of Theorem 2.

From Theorem 6 we see that all projective modules, and hence all free modules, are flat. It will appear presently that the class of flat modules can be strictly larger than the class of projective modules, but first we exhibit a non-flat module.

Exercise 7. *Show that if the ideal I contains a non-zero divisor and I is different from R, then R/I is a non-flat R-module.*

The next two exercises concern straightforward properties of flat modules.

Exercise 8. *Show that an arbitrary direct sum of flat R-modules is a flat R-module.*

Exercise 9. *Show that if $M_1, M_2, \dots, M_q$ are flat R-modules, then $M_1 \otimes M_2 \otimes \cdots \otimes M_q$ is a flat R-module.*

A deeper result is contained in

Exercise 10*. *Let M be an R-module with the property that every finitely generated submodule is contained in a flat submodule. Show that M itself is a flat R-module.*

It is now a simple matter to give an example of a flat module which is not projective. Such an example is provided by

Exercise 11*. *Show that the field $\mathbb{Q}$ of rational numbers, when considered as a $\mathbb{Z}$-module, is flat but not projective.*

The remaining comments on Chapter 2 all have to do with covariant extension. We already know from Theorem 8 and its corollary that projective modules stay projective and free modules stay free when they undergo such an extension. As the next exercise shows, a similar observation applies to flat modules.

Exercise 12. *Let $R \to S$ be a homomorphism of commutative rings and let M be a flat R-module. Show that $M \otimes_R S$ is a flat S-module.*

The main text makes a brief mention of different ways in which covariant extensions can arise, but so far we have not examined in detail any special cases. This omission will now be put right.

The simplest situation occurs when we have an ideal I of R and we use the natural ring-homomorphism $R \to R/I$. For an R-module M the corresponding covariant extension is $M \otimes_R (R/I)$ and this we know is naturally isomorphic to the R/I-module M/IM (see Exercise 3). The reader will find it an instructive exercise to check that the isomorphism described in Chapter 1, Exercise 5 can be regarded as a special case of Theorem 9.

Next let $X_1, X_2, \dots, X_n$ be indeterminates. Then R is a subring of the polynomial ring $R[X_1, X_2, \dots, X_n]$ and covariant extension by means of the inclusion homomorphism is referred to as *adjunction of the indeterminates* $X_1, X_2, \dots, X_n$. This can be looked at in a more down-to-earth manner.

Let M be an R-module and consider formal polynomials

$$\sum m_{\nu_1\nu_2\ldots\nu_n} X_1^{\nu_1} X_2^{\nu_2} \ldots X_n^{\nu_n},$$

where the coefficients $m_{\nu_1\nu_2\ldots\nu_n}$ are taken from M. It is clear that we can add such polynomials and that in this way we obtain an abelian group. Indeed if we denote the group by $M[X_1, X_2, \ldots, X_n]$, then $M[X_1, X_2, \ldots, X_n]$ can be regarded as an $R[X_1, X_2, \ldots, X_n]$-module in which

$$(rX_1^{\mu_1}X_2^{\mu_2} \ldots X_n^{\mu_n})(mX_1^{\nu_1}X_2^{\nu_2} \ldots X_n^{\nu_n}) = rmX_1^{\mu_1+\nu_1}X_2^{\mu_2+\nu_2} \ldots X_n^{\mu_n+\nu_n}.$$

(Here, of course, $r \in R$ and $m \in M$.) It is now a straightforward exercise to verify that there is an isomorphism

$$M[X_1, X_2, \ldots, X_n] \approx M \otimes_R R[X_1, X_2, \ldots, X_n]$$

of $R[X_1, X_2, \ldots, X_n]$-modules in which $mX_1^{\nu_1}X_2^{\nu_2} \ldots X_n^{\nu_n}$ is matched with $m \otimes X_1^{\nu_1}X_2^{\nu_2} \ldots X_n^{\nu_n}$. Note that $R[X_1, X_2, \ldots, X_n]$ is a free and therefore flat R-module.

The final example is taken from the general theory of fractions. First we recall how fractions are formed from the ring R.

Let Σ be a non-empty *multiplicatively closed* subset of R, that is to say Σ has the property that if $\sigma_1, \sigma_2, \ldots, \sigma_n$ $(n \geq 0)$ belong to it, then $\sigma_1\sigma_2 \ldots \sigma_n$ is in Σ as well. Note that in allowing $n=0$ as a possibility we are in fact assuming that the identity element of R is a member of the multiplicatively closed set.

We now consider formal fractions r/σ, where the numerator r is in R and the denominator σ is in Σ. Two such fractions, say r_1/σ_1 and r_2/σ_2, are regarded as the same and we write $r_1/\sigma_1 = r_2/\sigma_2$ if (and only if) $\sigma\sigma_2 r_1 = \sigma\sigma_1 r_2$ for some $\sigma \in \Sigma$; in other words we introduce an equivalence relation but without using a special notation for the equivalence classes. It is then possible to define addition and multiplication of fractions (compatible with the definition of equality) so that

$$\frac{r}{\sigma} + \frac{r'}{\sigma'} = \frac{\sigma' r + \sigma r'}{\sigma\sigma'}$$

and

$$\frac{r}{\sigma}\frac{r'}{\sigma'} = \frac{rr'}{\sigma\sigma'},$$

after which it is readily checked that the fractions form a commutative ring. This ring is known as the *ring of fractions* of R with respect to Σ and in what follows it will be denoted by $\Sigma^{-1}(R)$. Note that the identity element of $\Sigma^{-1}(R)$ is $1/1$, and that the fraction r/σ is the zero element of $\Sigma^{-1}(R)$ if and only if $\sigma' r = 0$ for some $\sigma' \in \Sigma$; also $\Sigma^{-1}(R)$ is a trivial ring only when 0 is a member of Σ.

In general R is not a subring of $\Sigma^{-1}(R)$ but we do have the canonical ring-homomorphism

$$R \to \Sigma^{-1}(R) \tag{2.6.5}$$

in which r of R is mapped into the fraction $r/1$. Let us examine what happens when we use (2.6.5) to pass from modules over R to modules over $\Sigma^{-1}(R)$.

To this end suppose that M is an R-module and let $x \in M \otimes_R \Sigma^{-1}(R)$. Then $x = m_1 \otimes \lambda_1 + m_2 \otimes \lambda_2 + \cdots + m_s \otimes \lambda_s$, where $m_i \in M$ and $\lambda_i \in \Sigma^{-1}(R)$. Now we can write the fractions λ_i with a common denominator in Σ, say $\lambda_i = r_i/\sigma$, and then

$$x = \sum_{i=1}^{s} m_i \otimes \frac{r_i}{\sigma} = \left(\sum_{i=1}^{s} r_i m_i\right) \otimes \frac{1}{\sigma}.$$

Thus each element of $M \otimes_R \Sigma^{-1}(R)$ can be written in the form $m \otimes (1/\sigma)$, where $m \in M$ and $\sigma \in \Sigma$.

Exercise 13*. *Let m belong to the R-module M and let σ belong to the multiplicatively closed subset Σ of R. Show that $m \otimes (1/\sigma) = 0$ in $M \otimes_R \Sigma^{-1}(R)$ if and only if $\sigma' m = 0$ for some $\sigma' \in \Sigma$.*

The construction which produced the ring $\Sigma^{-1}(R)$ from R can also be applied directly to an R-module M. Thus we may consider fractions of the form m/σ, where $m \in M$ and $\sigma \in \Sigma$; and naturally we treat m_1/σ_1 and m_2/σ_2 as the same if $\sigma\sigma_2 m_1 = \sigma\sigma_1 m_2$ for some $\sigma \in \Sigma$. These new fractions may be 'added', the sum of m/σ and m'/σ' being given by

$$\frac{m}{\sigma} + \frac{m'}{\sigma'} = \frac{\sigma' m + \sigma m'}{\sigma\sigma'}.$$

In this way we obtain an abelian group which is frequently denoted by $\Sigma^{-1}(M)$. In fact we can go further and regard $\Sigma^{-1}(M)$ as a $\Sigma^{-1}(R)$-module with

$$\frac{r}{\sigma}\left(\frac{m}{\sigma'}\right) = \frac{rm}{\sigma\sigma'}.$$

Finally, using Exercise 13, we can readily verify that there is an isomorphism

$$\Sigma^{-1}(M) \approx M \otimes_R \Sigma^{-1}(R)$$

of $\Sigma^{-1}(R)$-modules in which m/σ of $\Sigma^{-1}(M)$ corresponds to $m \otimes (1/\sigma)$. We leave to the reader the task of filling in the details.

Exercise 14*. *Let Σ be a multiplicatively closed subset of R and let $\Sigma^{-1}(R)$ be regarded as an R-module by means of the canonical ring-homomorphism $R \to \Sigma^{-1}(R)$. Show that $\Sigma^{-1}(R)$ is a flat R-module.*

We end these comments by showing that covariant extension is a transitive operation.

Suppose that $\omega: R \to S$ and $\lambda: S \to T$ are homomorphisms of commutative rings. By Theorem 7, if M is an R-module we have an (R, S)-isomorphism $(M \otimes_R S) \otimes_S T \approx M \otimes_R (S \otimes_S T)$ in which (with a self-explanatory notation) the element $(m \otimes s) \otimes t$ corresponds to $m \otimes (s \otimes t)$. On the other hand, Theorem 3 provides us with an isomorphism $S \otimes_S T \approx T$. Together these yield an isomorphism

$$(M \otimes_R S) \otimes_S T \approx M \otimes_R T \tag{2.6.6}$$

in which $(m \otimes s) \otimes t$ is matched with $m \otimes st$. To begin with (2.6.6) is an isomorphism of R-modules but, because of the way it operates, we can go further and say that it is actually an isomorphism of T-modules. Consequently if we extend M by means of ω and then extend the result by means of λ the total effect is essentially the same as if we had extended M by means of $\lambda \circ \omega$. This is what was meant by saying that covariant extension is a transitive operation.

2.7 Solutions to selected exercises

Exercise 2. *Let $x_1, x_2, \ldots, x_s$ be elements of an R-module M and $y_1, y_2, \ldots, y_s$ belong to an R-module N. Suppose that $x_1 \otimes y_1 + x_2 \otimes y_2 + \cdots + x_s \otimes y_s = 0$ in $M \otimes N$. Show that there exist finitely generated submodules M' and N', of M and N respectively, such that*

(i) *the x_i are in M' and the y_i are in N',*

and

(ii) *$x_1 \otimes y_1 + x_2 \otimes y_2 + \cdots + x_s \otimes y_s = 0$ in $M' \otimes N'$.*

Solution. Let $U(M, N)$ be the free R-module generated by $M \times N$ and let $V(M, N)$ be the submodule of $U(M, N)$ generated by elements having one or other of the following forms:

$$(m_1 + m_2, n) - (m_1, n) - (m_2, n),$$
$$(m, n_1 + n_2) - (m, n_1) - (m, n_2),$$
$$(rm, n) - r(m, n),$$
$$(m, rn) - r(m, n),$$

where the notation is self-explanatory. Now in the course of solving Problem 1 of Chapter 1 it was shown that $M \otimes N = U(M, N)/V(M, N)$; moreover in this context $m \otimes n$ is the natural image of the basis element (m, n) in $U(M, N)/V(M, N)$.

Next, because $x_1 \otimes y_1 + \cdots + x_s \otimes y_s = 0$ in $M \otimes N$, we see that $(x_1, y_1) + \cdots + (x_s, y_s)$ belongs to $V(M, N)$, that is to say

$$(x_1, y_1) + (x_2, y_2) + \cdots + (x_s, y_s) = r_1\xi_1 + r_2\xi_2 + \cdots + r_t\xi_t, \tag{2.7.1}$$

where $r_i \in R$ and ξ_i is an element of $U(M, N)$ having one or other of the four forms listed above. Note that (2.7.1) means simply that for each (m, n) in $M \times N$ the sum of the coefficients of all the occurrences of (m, n) in the left-hand side equals the sum of the coefficients of its occurrences in the right-hand side. But only finitely many basis elements (m, n) are actually present in (2.7.1) and therefore we can choose finitely generated submodules M' and N', of M and N respectively, so that if (m, n) appears on either side of (2.7.1), then $(m, n) \in M' \times N'$. This ensures that $x_1, x_2, \ldots, x_s$ are in M' and that $y_1, y_2, \ldots, y_s$ are in N'; it also ensures that $(x_1, y_1) + \cdots + (x_s, y_s)$ is in $V(M', N')$ as well as in $U(M', N')$. But $M' \otimes N' = U(M', N')/V(M', N')$ and therefore

$$x_1 \otimes y_1 + x_2 \otimes y_2 + \cdots + x_s \otimes y_s = 0$$

in $M' \otimes N'$ as required.

Exercise 3. *Let I be an ideal of R and let M be an R-module. Establish an isomorphism $M \otimes_R (R/I) \approx M/IM$ of R-modules.*

Solution. Let $\phi: R \to R/I$ be the natural homomorphism. Because tensor products are right exact, the exact sequence

$$0 \longrightarrow I \longrightarrow R \xrightarrow{\phi} R/I \longrightarrow 0$$

gives rise to the exact sequence

$$M \otimes I \to M \otimes R \to M \otimes (R/I) \to 0.$$

But $M \otimes R$ is isomorphic to M under an isomorphism which matches $m \otimes r$ with rm (see Theorem 3). It follows that there is a surjective homomorphism $M \to M \otimes (R/I)$ which maps m into $m \otimes \phi(1)$, and furthermore the kernel of this homomorphism is the image of $I \otimes M$ in M under the result of combining the homomorphisms

$$I \otimes M \longrightarrow R \otimes M \xrightarrow{\sim} M.$$

Accordingly the kernel of the homomorphism $M \to M \otimes (R/I)$ is IM and thus there is induced an isomorphism $M/IM \approx M \otimes (R/I)$.

Exercise 4. *Let M be a finitely generated R-module. Show that if $M = JM$ for every maximal ideal J, then $M = 0$.*

Solution. Let $m_1, m_2, \ldots, m_n$ generate M as an R-module and let J be a maximal ideal. Then, since m_i belongs to $M = JM$, we have

$$m_i = a_{i1}m_1 + a_{i2}m_2 + \cdots + a_{in}m_n$$

or

$$a_{i1}m_1 + a_{i2}m_2 + \cdots + (a_{ii} - 1)m_i + \cdots + a_{in}m_n = 0,$$

where the a_{ij} are in J. Consequently if D is the determinant

$$\begin{vmatrix} a_{11}-1 & a_{12} & \cdots & a_{1n} \\ a_{21} & a_{22}-1 & \cdots & a_{2n} \\ \vdots & \vdots & \vdots\vdots\vdots & \vdots \\ a_{n1} & a_{n2} & \cdots & a_{nn}-1 \end{vmatrix}$$

then $Dm_i=0$ for $i=1,2,\ldots,n$. It follows that D belongs to the ideal, $\mathrm{Ann}_R(M)$ say, formed by the elements of R that annihilate M. But $D=\alpha+(-1)^n$, where $\alpha\in J$, and therefore $D\notin J$. This shows that $\mathrm{Ann}_R(M)\nsubseteq J$, i.e. it shows that $\mathrm{Ann}_R(M)$ is not contained in any maximal ideal. Accordingly $\mathrm{Ann}_R(M)=R$ and now we see that $M=0$.

Exercise 5. *Let M be an R-module. Show that the following two statements are equivalent:*

(i) *for a finitely generated R-module N we have $M\otimes N=0$ only when $N=0$;*

(ii) *for every maximal ideal J, of R, JM is different from M.*

Solution. Assume (i) and let J be a maximal ideal. Then, since R/J is singly generated and non-zero, we have $M\otimes(R/J)\neq 0$. But, by Exercise 3, $M\otimes(R/J)$ is isomorphic to M/JM and so $JM\neq M$. Accordingly (i) implies (ii).

Assume (ii) and let N be a finitely generated R-module such that $M\otimes_R N=0$. If now J is a maximal ideal, then (Chapter 1, Exercise 5) $(M/JM)\otimes_{R/J}(N/JN)\approx(M\otimes_R N)/J(M\otimes_R N)=0$. But R/J is a field and M/JM and N/JN are vector spaces over R/J and therefore they are *free* R/J-modules. Furthermore, by hypothesis $M/JM\neq 0$. It follows (Chapter 1, Theorem 3) that $N/JN=0$. We now know that $N=JN$ for every maximal ideal J and so we may conclude, by virtue of Exercise 4, that $N=0$. This completes the solution.

Exercise 10. *Let M be an R-module with the property that every finitely generated submodule is contained in a flat submodule. Show that M itself is a flat R-module.*

Solution. Let $f: U\to V$ be an injective homomorphism of R-modules. It will be enough to show that $f\otimes M: U\otimes M\to V\otimes M$ is an injection as well. To this end let $\xi=u_1\otimes m_1+u_2\otimes m_2+\cdots+u_s\otimes m_s$, where $u_i\in U$ and $m_i\in M$, belong to the kernel of $f\otimes M$. Then $f(u_1)\otimes m_1+\cdots+f(u_s)\otimes m_s=0$ in $V\otimes M$. It now follows, from Exercise 2, that there exists a finitely generated submodule V^* (of V) containing $f(u_1), f(u_2),\ldots,f(u_s)$ and a finitely generated submodule M^* (of M) containing $m_1, m_2,\ldots,m_s$ such that

$$f(u_1) \otimes m_1 + f(u_2) \otimes m_2 + \cdots + f(u_s) \otimes m_s = 0$$

in $V^* \otimes M^*$. Choose a flat submodule M' of M so that $M^* \subseteq M'$. The obvious homomorphism $V^* \otimes M^* \to V \otimes M'$ then shows that $f(u_1) \otimes m_1 + \cdots + f(u_s) \otimes m_s = 0$ in $V \otimes M'$.

Let $u_1 \otimes m_1 + u_2 \otimes m_2 + \cdots + u_s \otimes m_s = x$ in $U \otimes M'$ and consider the natural commutative diagram

$$\begin{array}{ccc} U \otimes M' & \xrightarrow{f \otimes M'} & V \otimes M' \\ \downarrow & & \downarrow \\ U \otimes M & \xrightarrow{f \otimes M} & V \otimes M. \end{array}$$

The image of x in $V \otimes M'$ is zero, and $f \otimes M'$ is an injection because f is an injection and M' is flat. It follows that $x = 0$. But ξ is the image of x under the homomorphism $U \otimes M' \to U \otimes M$ so $\xi = 0$ as well.

Exercise 11. *Show that the field $\mathbb{Q}$ of rational numbers, when considered as a $\mathbb{Z}$-module, is flat but not projective.*

Solution. Let N be a finitely generated $\mathbb{Z}$-submodule of $\mathbb{Q}$. Then there exists $k \in \mathbb{Z}$, $k \neq 0$ such that N is contained in the $\mathbb{Z}$-submodule of $\mathbb{Q}$ generated by $1/k$. But the $\mathbb{Z}$-submodule generated by $1/k$ is isomorphic to $\mathbb{Z}$ and therefore N is isomorphic to an ideal of $\mathbb{Z}$. It follows that N is a free $\mathbb{Z}$-module so *a fortiori* it is $\mathbb{Z}$-flat. Exercise 10 now shows that $\mathbb{Q}$ is a flat $\mathbb{Z}$-module.

We next show that $\mathbb{Q}$ is not a submodule of a free $\mathbb{Z}$-module. (This is more than enough to establish that $\mathbb{Q}$ is not projective.) Assume the contrary and let $\mathbb{Q}$ be a submodule of a free $\mathbb{Z}$-module F with a base $\{\xi_i\}_{i \in I}$. Then we have a relation $1 = \sum m_i \xi_i$, where $m_i \in \mathbb{Z}$ and only finitely many m_i's are non-zero. Choose $j \in I$ so that $m_j \neq 0$ and then choose $p \in \mathbb{Z}$, $p \neq 0$ so that p does not divide m_j. Now

$$\frac{1}{p} = \sum n_i \xi_i,$$

where the n_i are integers, and it follows from this that $pn_j = m_j$ and therefore p divides m_j. This is the desired contradiction.

Exercise 13. *Let m belong to the R-module M and let σ belong to the multiplicatively closed subset Σ of R. Show that $m \otimes (1/\sigma) = 0$ in $M \otimes_R \Sigma^{-1}(R)$ if and only if $\sigma' m = 0$ for some $\sigma' \in \Sigma$.*

Solution. First suppose that $m \otimes (1/\sigma) = 0$ in $M \otimes_R \Sigma^{-1}(R)$. By Exercise 2, there exist finitely generated submodules M' and L', of M and $\Sigma^{-1}(R)$

respectively, such that $m \otimes (1/\sigma)=0$ in $M' \otimes L'$. Choose $\sigma_1 \in \Sigma$ so that $L' \subseteq R(1/\sigma\sigma_1)$; it then follows that $m \otimes (1/\sigma)=0$ in $M \otimes R(1/\sigma\sigma_1)$.

Consider the exact sequence

$$0 \longrightarrow I \longrightarrow R \xrightarrow{\phi} R\frac{1}{\sigma\sigma_1} \longrightarrow 0$$

of R-modules, where $\phi(r)=r/\sigma\sigma_1$. First we note that I consists of all the elements r of R such that $\sigma' r=0$ for some $\sigma' \in \Sigma$. Next there is induced an exact sequence

$$M \otimes I \longrightarrow M \otimes R \xrightarrow{M \otimes \phi} M \otimes R\frac{1}{\sigma\sigma_1} \longrightarrow 0$$

and, moreover, $m \otimes \sigma_1$ belongs to the kernel of $M \otimes \phi$, that is to say it belongs to the image of $M \otimes I$ in $M \otimes R$. The isomorphism $M \otimes R \approx M$ in which $m \otimes r$ is matched with rm now shows that $\sigma_1 m$ belongs to IM. We can therefore find $\sigma_2 \in \Sigma$ so that $\sigma_2\sigma_1 m=0$. But $\sigma_2\sigma_1 \in \Sigma$ so m is annihilated by an element of Σ.

Conversely suppose that $\sigma' m=0$ for some $\sigma' \in \Sigma$. Then in $M \otimes_R \Sigma^{-1}(R)$ we have

$$m \otimes \frac{1}{\sigma}=m \otimes \frac{\sigma'}{\sigma\sigma'}=\sigma' m \otimes \frac{1}{\sigma\sigma'}=0$$

as required.

Exercise 14. *Let Σ be a multiplicatively closed subset of R and let $\Sigma^{-1}(R)$ be regarded as an R-module by means of the canonical ring-homomorphism $R \to \Sigma^{-1}(R)$. Show that $\Sigma^{-1}(R)$ is a flat R-module.*

Solution. Let $f: M' \to M$ be an injective homomorphism of R-modules. It is sufficient to show that $f \otimes \Sigma^{-1}(R)$ is also an injection. To this end let x in $M' \otimes \Sigma^{-1}(R)$ belong to the kernel of $f \otimes \Sigma^{-1}(R)$, and let $x=m' \otimes (1/\sigma)$, where $m' \in M'$ and $\sigma \in \Sigma$. (Every element of $M' \otimes_R \Sigma^{-1}(R)$ can be written in this form.) Then $f(m') \otimes (1/\sigma)=0$ and therefore, by Exercise 13, $\sigma' f(m')=0$ for some $\sigma' \in \Sigma$. Accordingly $f(\sigma' m')=0$ and therefore $\sigma' m'=0$ because f is an injection. Finally

$$x=m' \otimes \frac{1}{\sigma}=m' \otimes \frac{\sigma'}{\sigma\sigma'}=\sigma' m' \otimes \frac{1}{\sigma\sigma'}=0$$

and the solution is complete.

3

Associative algebras

General remarks

Our main objective is to discuss the important algebras that can be derived from a module, and so in this, the last of the introductory chapters, we shall study those general aspects of the theory of Algebras that are relevant to our goal. At the outset it needs to be stressed that we shall be concerned solely with algebras that are associative and possess an identity element. Accordingly, from now on, the term algebra will only be used in this restricted sense.

As before R and S denote commutative rings, each with an identity element, and a ring-homomorphism between them is required to take identity element into identity element. Later, when we come to consider homomorphisms of algebras, these too will be required to preserve identity elements.

Finally when dealing with tensor products of modules we shall sometimes omit from the symbol $\otimes$ the suffix which indicates the underlying ring. In this chapter whenever the suffix is left out the ring in question is always R.

3.1 Basic definitions

Let A be a (not necessarily commutative) ring with an identity element 1_A, and at the same time let it be a module over the commutative ring R. We suppose that the sum of two elements of A is the same whether we use the ring or the module structure. In these circumstances A is called an *R-algebra* provided that

$$r(a_1a_2)=(ra_1)a_2=a_1(ra_2) \tag{3.1.1}$$

whenever $a_1, a_2 \in A$ and $r \in R$. This condition is equivalent to

$$(r_1a_1)(r_2a_2)=(r_1r_2)(a_1a_2), \tag{3.1.2}$$

where now r_1, r_2 denote elements of R.

Let A be an R-algebra. The mapping

$$\phi: R \to A, \tag{3.1.3}$$

defined by $\phi(r) = r1_A$, is both a ring-homomorphism and a homomorphism of R-modules. Furthermore, since $\phi(r)a = ra = a\phi(r)$, $\phi(R)$ is contained in the *centre* of A. We shall call (3.1.3) the *structural homomorphism* of the R-algebra.

This last concept provides an alternative way of looking at R-algebras. Suppose that A is a ring with identity element and assume that we are given a ring-homomorphism $\phi: R \to A$ which preserves identity elements and maps R into the centre of A. If we turn A into an R-module by putting $ra = \phi(r)a$, then A becomes an R-algebra with ϕ as its structural homomorphism. For example, we see that R itself is an R-algebra with the identity mapping as the structural homomorphism.

An important example of an R-algebra arises in the following way. Let M and N be R-modules. The R-linear mappings of M into N can be added and they can be multiplied by elements of R. In fact they form the R-module which is denoted by $\mathrm{Hom}_R(M, N)$. When $N = M$ these homomorphisms are the *endomorphisms* of M and in this case we use the notation $\mathrm{End}_R(M)$ rather than $\mathrm{Hom}_R(M, M)$. Now if f, g belong to $\mathrm{End}_R(M)$, then so does $f \circ g$. Indeed it is easy to check that $\mathrm{End}_R(M)$ becomes a ring with identity if the product of f and g is taken to be $f \circ g$. But for $r \in R$ we have

$$(rf) \circ g = r(f \circ g) = f \circ (rg),$$

so that in fact $\mathrm{End}_R(M)$ is an R-algebra. We call $\mathrm{End}_R(M)$ the *endomorphism algebra* of M. The identity element of $\mathrm{End}_R(M)$ is the identity mapping of M and in the structural homomorphism $R \to \mathrm{End}_R(M)$ an element r, of R, is mapped into the corresponding *homothety* that is the mapping $M \to M$ in which $m \in M$ goes into rm.

We shall now introduce some further terminology. Let A and B be R-algebras. A mapping $f: A \to B$ is called an *algebra-homomorphism*, or a *homomorphism of R-algebras*, if it is both a homomorphism of rings (preserving identity elements) and a homomorphism of R-modules. For example the structural homomorphism $R \to A$ is a homomorphism of R-algebras. Note that if $\phi: R \to A$ and $\psi: R \to B$ are the structural homomorphisms of A and B, then a mapping $f: A \to B$ is an algebra-homomorphism if and only if it is a homomorphism of rings and $f \circ \phi = \psi$. Naturally a bijective algebra-homomorphism is called an *algebra-isomorphism*.

Suppose next that C is a subring of the R-algebra A (with $1_C = 1_A$) and suppose that it is an R-submodule of A as well. Then C itself is an R-algebra. Such an algebra is called a *subalgebra*, or an *R-subalgebra*, of A. The intersection of any family of subalgebras of A is again a subalgebra. Hence if U is any *subset* of A, there will be a smallest subalgebra, B say, containing U. We call B the subalgebra *generated by* U. As an R-module B is spanned by all products $u_1 u_2 \ldots u_n$, where $u_i \in U$ and $n \geq 0$. Here when $n=0$ we have to do with the *empty product* and this is interpreted as being the identity element of A. Of course the inclusion mapping into A of one of its subalgebras is an algebra-homomorphism.

Finally suppose that I is a two-sided ideal of the R-algebra A. Then I is necessarily an R-submodule. Accordingly A/I is both a ring and an R-module. Indeed A/I is an R-algebra. Note that the natural mapping of A onto the factor algebra A/I is a homomorphism of R-algebras.

3.2 Tensor products of algebras

Let $A_1, A_2, \ldots, A_p$ be R-algebras. Then $A_1 \otimes_R A_2 \otimes_R \cdots \otimes_R A_p$ is certainly an R-module. Indeed, as will be established shortly, it has a natural structure as an R-algebra.

Consider the multilinear mapping

$$A_1 \times A_2 \times \cdots \times A_p \times A_1 \times A_2 \times \cdots \times A_p \to A_1 \otimes A_2 \otimes \cdots \otimes A_p$$

in which $(a_1, a_2, \ldots, a_p, a_1', a_2', \ldots, a_p')$ is mapped into $a_1 a_1' \otimes a_2 a_2' \otimes \cdots \otimes a_p a_p'$. This (Chapter 1, Theorem 2) induces a homomorphism

$$\begin{aligned} &A_1 \otimes A_2 \otimes \cdots \otimes A_p \otimes A_1 \otimes A_2 \otimes \cdots \otimes A_p \\ &\qquad \to A_1 \otimes A_2 \otimes \cdots \otimes A_p \end{aligned} \tag{3.2.1}$$

of R-modules. Moreover (Chapter 2, Theorem 1) we have an R-isomorphism

$$\begin{aligned} &(A_1 \otimes \cdots \otimes A_p) \otimes (A_1 \otimes \cdots \otimes A_p) \\ &\qquad \approx A_1 \otimes \cdots \otimes A_p \otimes A_1 \otimes \cdots \otimes A_p \end{aligned} \tag{3.2.2}$$

in which, with a self-explanatory notation, $(a_1 \otimes \cdots \otimes a_p) \otimes (a_1' \otimes \cdots \otimes a_p')$ corresponds to $a_1 \otimes \cdots \otimes a_p \otimes a_1' \otimes \cdots \otimes a_p'$. If therefore we combine (3.2.1) and (3.2.2) the result is a homomorphism

$$(A_1 \otimes \cdots \otimes A_p) \otimes (A_1 \otimes \cdots \otimes A_p) \to A_1 \otimes A_2 \otimes \cdots \otimes A_p \tag{3.2.3}$$

in which $(a_1 \otimes \cdots \otimes a_p) \otimes (a_1' \otimes \cdots \otimes a_p')$ is carried over into $a_1 a_1' \otimes a_2 a_2' \otimes \cdots \otimes a_p a_p'$.

We now define a mapping

$$\begin{aligned} \mu \colon &(A_1 \otimes \cdots \otimes A_p) \times (A_1 \otimes \cdots \otimes A_p) \\ &\qquad \to A_1 \otimes A_2 \otimes \cdots \otimes A_p \end{aligned} \tag{3.2.4}$$

by requiring $\mu(x, x')$, where x and x' belong to $A_1 \otimes A_2 \otimes \cdots \otimes A_p$, to be the image of $x \otimes x'$ under the mapping (3.2.3). Evidently μ is bilinear and

$$\mu(a_1 \otimes \cdots \otimes a_p, a_1' \otimes \cdots \otimes a_p')$$
$$= a_1 a_1' \otimes a_2 a_2' \otimes \cdots \otimes a_p a_p'. \quad (3.2.5)$$

It follows that if $x = a_1 \otimes \cdots \otimes a_p$, $x' = a_1' \otimes \cdots \otimes a_p'$, and $x'' = a_1'' \otimes \cdots \otimes a_p''$, then

$$\mu(\mu(x, x'), x'') = \mu(x, \mu(x', x'')). \quad (3.2.6)$$

But μ is bilinear. Consequently (3.2.6) holds when x, x' and x'' are *any* three elements of $A_1 \otimes A_2 \otimes \cdots \otimes A_p$.

We are now ready to turn $A_1 \otimes A_2 \otimes \cdots \otimes A_p$ into an R-algebra. To do this we define the product of two of its elements, say x and x', to be $\mu(x, x')$. It then follows from (3.2.6) that multiplication is associative; and the bilinearity of μ ensures that it is distributive with respect to addition. Let e_i be the identity element of A_i. By (3.2.5) we have

$$\mu(e_1 \otimes \cdots \otimes e_p, a_1 \otimes \cdots \otimes a_p)$$
$$= a_1 \otimes \cdots \otimes a_p = \mu(a_1 \otimes \cdots \otimes a_p, e_1 \otimes \cdots \otimes e_p)$$

and therefore

$$\mu(e_1 \otimes \cdots \otimes e_p, x) = x = \mu(x, e_1 \otimes \cdots \otimes e_p)$$

for all x in $A_1 \otimes A_2 \otimes \cdots \otimes A_p$. Thus $A_1 \otimes A_2 \otimes \cdots \otimes A_p$ is a ring with identity. Finally for $r \in R$ we have

$$\mu(rx, x') = r\mu(x, x') = \mu(x, rx')$$

and therefore $A_1 \otimes A_2 \otimes \cdots \otimes A_p$ is an R-algebra.

We summarize our conclusions so far.

Theorem 1. *Let $A_1, A_2, \ldots, A_p$ be R-algebras. Then*

$$A_1 \otimes_R A_2 \otimes_R \cdots \otimes_R A_p$$

is an R-algebra, where the R-module structure is the usual one and the product of $a_1 \otimes a_2 \otimes \cdots \otimes a_p$ and $a_1' \otimes a_2' \otimes \cdots \otimes a_p'$ is $a_1 a_1' \otimes a_2 a_2' \otimes \cdots \otimes a_p a_p'$.

It will now be shown that the results of Section (2.1) yield certain isomorphisms between algebras. We can deal with these fairly rapidly.

Theorem 2. *Let $A_1, A_2, \ldots, A_p$ and $B_1, B_2, \ldots, B_q$ be R-algebras. Then there is an isomorphism*

$$A_1 \otimes_R \cdots \otimes_R A_p \otimes B_1 \otimes_R \cdots \otimes_R B_q$$
$$\approx (A_1 \otimes_R \cdots \otimes_R A_p) \otimes_R (B_1 \otimes_R \cdots \otimes_R B_q)$$

of R-algebras in which $a_1 \otimes \cdots \otimes a_p \otimes b_1 \otimes \cdots \otimes b_q$ is associated with $(a_1 \otimes \cdots \otimes a_p) \otimes (b_1 \otimes \cdots \otimes b_q)$.

Proof. By Theorem 1 of Chapter 2, there is an isomorphism f, *of R-modules*, that maps $A_1 \otimes \cdots \otimes A_p \otimes B_1 \otimes \cdots \otimes B_q$ onto $(A_1 \otimes \cdots \otimes A_p) \otimes (B_1 \otimes \cdots \otimes B_q)$ so that

$$f(a_1 \otimes \cdots \otimes a_p \otimes b_1 \otimes \cdots \otimes b_q)$$
$$= (a_1 \otimes \cdots \otimes a_p) \otimes (b_1 \otimes \cdots \otimes b_q).$$

Let $x = a_1 \otimes \cdots \otimes a_p \otimes b_1 \otimes \cdots \otimes b_q$ and $x' = a'_1 \otimes \cdots \otimes a'_p \otimes b'_1 \otimes \cdots \otimes b'_q$. Then, by Theorem 1,

$$xx' = a_1 a'_1 \otimes \cdots \otimes a_p a'_p \otimes b_1 b'_1 \otimes \cdots \otimes b_q b'_q$$

and therefore

$$f(xx') = (a_1 a'_1 \otimes \cdots \otimes a_p a'_p) \otimes (b_1 b'_1 \otimes \cdots \otimes b_q b'_q)$$
$$= f(x) f(x').$$

But f is R-linear. Hence if y and y' are *any* two elements of $A_1 \otimes \cdots \otimes A_p \otimes B_1 \otimes \cdots \otimes B_q$, then $f(yy') = f(y)f(y')$. The theorem follows because f is bijective.

Before we leave this result we note that the argument used to deduce the corollary to Theorem 1 of Chapter 2 now shows that there is an isomorphism

$$(A_1 \otimes_R A_2) \otimes_R A_3 \approx A_1 \otimes_R (A_2 \otimes_R A_3) \tag{3.2.7}$$

of R-algebras in which $(a_1 \otimes a_2) \otimes a_3$ corresponds to $a_1 \otimes (a_2 \otimes a_3)$.

The next theorem is derived, in much the same way, from Theorem 2 of Chapter 2. The details are left to the reader.

Theorem 3. *Let $A_1, A_2, \ldots, A_p$ be R-algebras and let $(i_1, i_2, \ldots, i_p)$ be a permutation of $(1, 2, \ldots, p)$. Then there is an isomorphism*

$$A_1 \otimes_R A_2 \otimes_R \cdots \otimes_R A_p \approx A_{i_1} \otimes_R A_{i_2} \otimes_R \cdots \otimes_R A_{i_p}$$

of R-algebras in which $a_1 \otimes a_2 \otimes \cdots \otimes a_p$ corresponds to $a_{i_1} \otimes a_{i_2} \otimes \cdots \otimes a_{i_p}$.

It will be recalled that R itself is an R-algebra.

Theorem 4. *Let A be an R-algebra. Then there is an isomorphism $R \otimes_R A \approx A$ of R-algebras which matches $r \otimes a$ with ra. There is a similar algebra-isomorphism $A \otimes_R R \approx A$, where $a \otimes r$ is matched with ra.*

Proof. We need only consider the first assertion. By Theorem 3 of Chapter 2, there is an isomorphism $f: R \otimes_R A \to A$ of R-modules for which $f(r \otimes a) = ra$. That f is compatible with multiplication is clear because $(r_1 a_1)(r_2 a_2) = (r_1 r_2)(a_1 a_2)$.

We next examine tensor products in relation to homomorphisms of algebras. To this end let $A_1, A_2, \ldots, A_p$ and $B_1, B_2, \ldots, B_p$ be R-algebras and suppose that we are given algebra-homomorphisms $f_i: A_i \to B_i$ for $i = 1, 2, \ldots, p$. From Section (2.2) we know that

$$f_1 \otimes f_2 \otimes \cdots \otimes f_p: A_1 \otimes_R A_2 \otimes_R \cdots \otimes_R A_p$$
$$\to B_1 \otimes_R B_2 \otimes_R \cdots \otimes_R B_p \tag{3.2.8}$$

is a homomorphism of R-modules. *We claim that in the present instance it is actually a homomorphism of R-algebras.* For it is clear that $f_1 \otimes f_2 \otimes \cdots \otimes f_p$ takes identity element into identity element. Now suppose that x and x' belong to $A_1 \otimes A_2 \otimes \cdots \otimes A_p$. It suffices to show that the image of xx' is the product of the separate images of x and x'. Indeed it is enough to prove that this is so when x and x' have the special forms $a_1 \otimes a_2 \otimes \cdots \otimes a_p$ and $a'_1 \otimes a'_2 \otimes \cdots \otimes a'_p$. But in these circumstances what we wish to prove is obvious.

3.3 Graded algebras

Let A be an R-algebra and $\{A_n\}_{n \in \mathbb{Z}}$ a family of R-submodules of A indexed by the integers. The family is said to constitute a *grading* on A provided that

(i) $A = \sum_{n \in \mathbb{Z}} A_n$ (d.s.),

and

(ii) $a_r \in A_r$ *and* $a_s \in A_s$ *together imply that* $a_r a_s \in A_{r+s}$.

An R-algebra with a grading is known as a *graded* R-algebra. The elements of A_n are said to be *homogeneous* of *degree* n. Thus condition (ii) says that the product of two homogeneous elements is again homogeneous, and the degree of the product is the sum of the degrees of the two factors.

We shall frequently meet with situations where $A_n = 0$ for all $n < 0$. A grading with this additional property will be called a *non-negative* grading.

Suppose that A is graded by $\{A_n\}_{n \in \mathbb{Z}}$. Then an element $a \in A$ has a unique representation in the form

$$a = \cdots + a_{-2} + a_{-1} + a_0 + a_1 + a_2 + \cdots,$$

where $a_n \in A_n$ and only finitely many terms on the right-hand side are non-zero. We call the summands in the representation the *homogeneous components* of a.

Theorem 5. *Let* $\{A_n\}_{n \in \mathbb{Z}}$ *be a grading on the R-algebra A. Then the identity element* 1_A *belongs to* A_0.

Proof. Let ε be the homogeneous component of 1_A of degree zero and let $a_n \in A_n$. By comparing the components of degree n in the equations $1_A a_n = a_n = a_n 1_A$ we see that $\varepsilon a_n = a_n = a_n \varepsilon$. But every element of A is a sum of homogeneous elements and therefore $\varepsilon x = x = x\varepsilon$ for all $x \in A$. Accordingly $1_A = \varepsilon$ and the theorem is proved.

Corollary. A_0 *is an R-subalgebra of A.*

An important simple situation arises when A is generated, as an R-

algebra, by its homogeneous elements of degree one. The relevant facts are recorded in the next lemma.

Lemma 1. *Let $\{A_n\}_{n\in\mathbb{Z}}$ be a grading on the R-algebra A, and suppose that A is generated, as an R-algebra, by A_1. Then the grading is non-negative, and, for $p\geq 1$, each element of A_p is a sum of products of p elements of A_1. Furthermore A_0 is generated, as an R-module, by the identity element 1_A and therefore A_0 is contained in the centre of A.*

Proof. All the assertions become obvious as soon as it is noted that (i) A is generated, *as an R-module*, by products of elements of A_1 (including the empty product), and (ii) such a product is homogeneous of degree p, where p is the number of factors.

We need some further terminology. Let A and B be graded R-algebras. A mapping $f: A\to B$ is called a *homomorphism of graded algebras* if it is a homomorphism of R-algebras which preserves degrees, i.e. if it satisfies $f(A_n)\subseteq B_n$ for every n. (Here $\{A_n\}_{n\in\mathbb{Z}}$ and $\{B_n\}_{n\in\mathbb{Z}}$ are the respective gradings.) By an *isomorphism of graded algebras* we naturally mean a bijective homomorphism of graded algebras. If $f: A\to B$ is such an isomorphism, then, for each n, f maps A_n isomorphically onto B_n.

Now let H be a two-sided ideal of a graded R-algebra A.

Definition. *The two-sided ideal H is called 'homogeneous' if whenever $a\in H$ all the homogeneous components of a belong to H as well.*

Lemma 2. *Let H be a two-sided ideal of a graded R-algebra A. Then H is homogeneous if and only if it can be generated by homogeneous elements.*

Proof. If H is homogeneous, then it is certainly generated by the homogeneous components of the various members of H. Thus H has a homogeneous system of generators.

Conversely suppose that Ω is a set of homogeneous elements and that Ω generates H. Then each element of H is a sum of elements of the form axb, where $a, b\in A$ and $x\in\Omega$. Indeed we may suppose that a and b are themselves homogeneous. Thus if $\alpha\in H$, then α is a sum of homogeneous elements of H and therefore the homogeneous components of α belong to H. This completes the proof.

Now suppose that H is a homogeneous two-sided ideal of the graded R-algebra A. Put

$$H_n = H\cap A_n. \tag{3.3.1}$$

Then

$$H=\sum_{n\in\mathbb{Z}} H_n \quad \text{(d.s.)}. \tag{3.3.2}$$

We already know that $A/H = B$ say is an R-algebra. Suppose that we put

$$B_n = (A_n + H)/H. \tag{3.3.3}$$

Then B_n is an R-submodule of B and

$$B = \sum_{n \in \mathbb{Z}} B_n. \tag{3.3.4}$$

We claim that the sum (3.3.4) *is direct*. To see this suppose that $\sum_n b_n = 0$, where $b_n \in B_n$ and there are only finitely many non-zero summands. For each $n \in \mathbb{Z}$ we choose $a_n \in A_n$ so that $b_n = a_n + H$, taking care to arrange that $a_n = 0$ whenever $b_n = 0$. Then $\sum_n a_n$ belongs to H and therefore, because H is homogeneous, all the a_n are in H. But this means that all the b_n are zero. Accordingly our claim is established.

Clearly the product of an element of B_r and an element of B_s is an element of B_{r+s}. Hence the R-algebra A/H is graded by its submodules $\{(A_n + H)/H\}_{n \in \mathbb{Z}}$ and the natural mapping $A \to A/H$ is a homomorphism of graded algebras. This mapping induces, for each n, a homomorphism of A_n onto B_n whose kernel is $A_n \cap H = H_n$. Thus we have isomorphisms

$$A_n/H_n \approx (A_n + H)/H \tag{3.3.5}$$

of R-modules and these can be combined to give an isomorphism

$$A/H \approx \sum_{n \in \mathbb{Z}} A_n/H_n \quad \text{(d.s.)}. \tag{3.3.6}$$

In the first instance (3.3.6) is just an isomorphism of R-modules, but of course there is a unique way to turn the right-hand side into a graded R-algebra so that (3.3.6) becomes an isomorphism of graded algebras.

We next investigate the tensor product of $A^{(1)}, A^{(2)}, \ldots, A^{(p)}$, where each $A^{(i)}$ is a graded R-algebra. By Theorem 5 of Chapter 2,

$$A^{(1)} \otimes A^{(2)} \otimes \cdots \otimes A^{(p)} = \sum A^{(1)}_{i_1} \otimes A^{(2)}_{i_2} \otimes \cdots \otimes A^{(p)}_{i_p} \quad \text{(d.s.)},$$

where the direct sum is taken over all sequences $(i_1, i_2, \ldots, i_p)$ of p integers. Put

$$A = A^{(1)} \otimes A^{(2)} \otimes \cdots \otimes A^{(p)} \tag{3.3.7}$$

and for each sequence $I = (i_1, i_2, \ldots, i_p)$ let

$$A_I = A^{(1)}_{i_1} \otimes A^{(2)}_{i_2} \otimes \cdots \otimes A^{(p)}_{i_p}. \tag{3.3.8}$$

Then (to recapitulate) A is an R-algebra and

$$A = \sum_I A_I \quad \text{(d.s.)} \tag{3.3.9}$$

this being a direct sum of R-modules. Furthermore if $J = (j_1, j_2, \ldots, j_p)$ is a second sequence of integers, then for $x \in A_I$ and $y \in A_J$ we have $xy \in A_{I+J}$, where $I + J = (i_1 + j_1, i_2 + j_2, \ldots, i_p + j_p)$.

It will be convenient to put

$$|I| = i_1 + i_2 + \cdots + i_p \tag{3.3.10}$$

and

$$A_n = \sum_{|I|=n} A_I = \sum_{|I|=n} A^{(1)}_{i_1} \otimes_R A^{(2)}_{i_2} \otimes_R \cdots \otimes_R A^{(p)}_{i_p}. \tag{3.3.11}$$

Then as a consequence of (3.3.9) we have

$$A = \sum_{n \in \mathbb{Z}} A_n \quad \text{(d.s.)} \tag{3.3.12}$$

and it is clear that the product of an element of A_n and element of A_k belongs to A_{n+k}.

Let us sum up. It has now been shown that if $A^{(1)}, A^{(2)}, \ldots, A^{(p)}$ are graded R-algebras, then $A^{(1)} \otimes_R A^{(2)} \otimes_R \cdots \otimes_R A^{(p)}$ is also a graded R-algebra, where the module of homogeneous elements of degree n is

$$A_n = \sum_{|I|=n} A^{(1)}_{i_1} \otimes_R A^{(2)}_{i_2} \otimes_R \cdots \otimes_R A^{(p)}_{i_p}.$$

We call $\{A_n\}_{n \in \mathbb{Z}}$ the *total grading* on the tensor product. Evidently if the gradings on $A^{(1)}, A^{(2)}, \ldots, A^{(p)}$ are non-negative, then the total grading on $A^{(1)} \otimes A^{(2)} \otimes \cdots \otimes A^{(p)}$ is non-negative as well.

We next review some of our basic isomorphisms from the standpoint of the theory of graded algebras. Suppose then that $A^{(1)}, A^{(2)}, \ldots, A^{(p)}$ and $B^{(1)}, B^{(2)}, \ldots, B^{(q)}$ are graded R-algebras. It will be recalled that Theorem 2 provides an explicit isomorphism

$$\begin{aligned} &A^{(1)} \otimes_R \cdots \otimes_R A^{(p)} \otimes_R B^{(1)} \otimes_R \cdots \otimes_R B^{(q)} \\ &\qquad \approx (A^{(1)} \otimes_R \cdots \otimes_R A^{(p)}) \otimes_R (B^{(1)} \otimes_R \cdots \otimes_R B^{(q)}) \end{aligned} \tag{3.3.13}$$

of ordinary, that is *ungraded*, R-algebras. However, on this occasion each side of (3.3.13) is a graded algebra and it is clear (from the way the isomorphism operates and the way the gradings are defined) that the degrees of homogeneous elements are preserved. Accordingly what we have is an isomorphism of *graded algebras*.

Let us consider Theorem 3 in a similar way. To this end suppose that $(\mu_1, \mu_2, \ldots, \mu_p)$ is a permutation of $(1, 2, \ldots, p)$. Then the isomorphism

$$\begin{aligned} &A^{(1)} \otimes_R A^{(2)} \otimes_R \cdots \otimes_R A^{(p)} \\ &\qquad \approx A^{(\mu_1)} \otimes_R A^{(\mu_2)} \otimes_R \cdots \otimes_R A^{(\mu_p)} \end{aligned} \tag{3.3.14}$$

that we get from Theorem 3 preserves degrees and so it too is an isomorphism of graded algebras.

In the case of Theorem 4 a little more explanation is needed. If B is an R-algebra we can obtain a grading on B by putting $B_0 = B$ and $B_n = 0$ whenever $n \neq 0$. This is known as the *trivial grading*. Suppose now that A is a graded R-algebra. Then Theorem 4 provides algebra-isomorphisms

$$R \otimes_R A \approx A \tag{3.3.15}$$

and

$$A \otimes_R R \approx A. \tag{3.3.16}$$

It is readily checked that these are isomorphisms of *graded algebras* provided that we endow R with the trivial grading.

Our final observation in this section concerns homomorphisms $f_i: A^{(i)} \to B^{(i)}$, where $i=1, 2, \ldots, p$, of graded algebras. It was observed in Section (3.2) that $f_1 \otimes f_2 \otimes \cdots \otimes f_p$ is a homomorphism

$$A^{(1)} \otimes_R A^{(2)} \otimes_R \cdots \otimes_R A^{(p)} \to B^{(1)} \otimes_R B^{(2)} \otimes_R \cdots \otimes_R B^{(p)}$$

of R-algebras. However, in the present situation it has the additional property of preserving degrees and therefore, this time, $f_1 \otimes f_2 \otimes \cdots \otimes f_p$ is a homomorphism of graded algebras.

3.4 A modified graded tensor product

Let $A^{(1)}, A^{(2)}, \ldots, A^{(p)}$ be graded R-algebras. We have shown that $A^{(1)} \otimes A^{(2)} \otimes \cdots \otimes A^{(p)} = A$ say has a natural structure as a graded R-algebra. We propose to modify this structure. The outcome will be that A will remain a graded R-algebra with the same R-module structure, the same grading and the same identity element. Only the definition of the product of two elements of A will be changed.

Let $I=(i_1, i_2, \ldots, i_p)$ be a sequence of p integers and put

$$A_I = A^{(1)}_{i_1} \otimes A^{(2)}_{i_2} \otimes \cdots \otimes A^{(p)}_{i_p}$$

as we did before. Then A is the direct sum of the A_I so that each $a \in A$ has a unique representation in the form

$$a = \sum_I a_I, \tag{3.4.1}$$

where $a_I \in A_I$ and only finitely many of the summands are non-zero.

Next let $J=(j_1, j_2, \ldots, j_p)$ be a second sequence of p integers. The multiplication already introduced on A provides a bilinear mapping

$$\mu_{IJ}: A_I \times A_J \to A_{I+J},$$

where $I+J=(i_1+j_1, i_2+j_2, \ldots, i_p+j_p)$. Put

$$N(I,J) = \sum_{r>s} i_r j_s \tag{3.4.2}$$

and

$$\varepsilon(I,J) = (-1)^{N(I,J)}. \tag{3.4.3}$$

We are now ready to describe the new or modified multiplication.

Suppose that

$$a=\sum_I a_I \quad \text{and} \quad a'=\sum_J a'_J$$

are elements of A. Then the *modified product* of a and a' is defined to be

$$\sum_{I,J} \varepsilon(I, J)\mu_{IJ}(a_I, a'_J). \tag{3.4.4}$$

Thus if $a_{i_1} \otimes a_{i_2} \otimes \cdots \otimes a_{i_p}$ belongs to A_I and $a'_{j_1} \otimes a'_{j_2} \otimes \cdots \otimes a'_{j_p}$ belongs to A_J, then, whereas their ordinary product is $a_{i_1}a'_{j_1} \otimes a_{i_2}a'_{j_2} \otimes \cdots \otimes a_{i_p}a'_{j_p}$, their modified product is

$$\varepsilon(I, J)a_{i_1}a'_{j_1} \otimes a_{i_2}a'_{j_2} \otimes \cdots \otimes a_{i_p}a'_{j_p}. \tag{3.4.5}$$

Clearly modified multiplication is distributive with respect to addition on both sides, and $e^{(1)} \otimes e^{(2)} \otimes \cdots \otimes e^{(p)}$, where $e^{(r)}$ is the identity element of $A^{(r)}$, is still neutral for the new multiplication. Also the property that corresponds to (3.1.1) continues to hold. Hence in order to check that we still have an R-algebra it suffices to verify that modified multiplication is associative.

Let $I=(i_1, i_2, \ldots, i_p)$, $J=(j_1, j_2, \ldots, j_p)$ and $K=(k_1, k_2, \ldots, k_p)$ be sequences of integers and suppose that $x \in A_I$, $y \in A_J$ and $z \in A_K$. It will suffice to verify the associative law in the case of x, y and z. Indeed we may go further and assume that $x=a_{i_1} \otimes \cdots \otimes a_{i_p}$, $y=a'_{j_1} \otimes \cdots \otimes a'_{j_p}$ and $z=a''_{k_1} \otimes \cdots \otimes a''_{k_p}$. But now, in view of (3.4.5) what we wish to establish will follow if we show that

$$\varepsilon(I, J)\varepsilon(I+J, K)=\varepsilon(I, J+K)\varepsilon(J, K).$$

However, this is clear because $\varepsilon(I+J, K)=\varepsilon(I, K)\varepsilon(J, K)$ and $\varepsilon(I, J+K)=\varepsilon(I, J)\varepsilon(I, K)$.

The new R-algebra will be called the *modified tensor product* of $A^{(1)}, A^{(2)}, \ldots, A^{(p)}$. It will be denoted by

$$A^{(1)} \underline{\otimes}_R A^{(2)} \underline{\otimes}_R \cdots \underline{\otimes}_R A^{(p)} \tag{3.4.6}$$

in order to distinguish it from $A^{(1)} \otimes A^{(2)} \otimes \cdots \otimes A^{(p)}$. Note that the total grading on $A^{(1)} \otimes A^{(2)} \otimes \cdots \otimes A^{(p)}$ also serves as a grading on $A^{(1)} \underline{\otimes} A^{(2)} \underline{\otimes} \cdots \underline{\otimes} A^{(p)}$. Finally we recall that the two tensor products have the same identity element; and that for $x \in A_I, y \in A_J$ the modified product of x and y is $\varepsilon(I, J)$ times the ordinary product.

The new tensor product is important in the theory of exterior algebras. In the remainder of this section we shall prepare the way for its application.

Theorem 6. *Let $A^{(1)}, A^{(2)}, \ldots, A^{(p)}$ and $B^{(1)}, B^{(2)}, \ldots, B^{(q)}$ be graded R-algebras. Then*

$$A^{(1)} \underline{\otimes}_R \cdots \underline{\otimes}_R A^{(p)} \underline{\otimes}_R B^{(1)} \underline{\otimes}_R \cdots \underline{\otimes}_R B^{(q)} \tag{3.4.7}$$

and

$$(A^{(1)} \underline{\otimes}_R \cdots \underline{\otimes}_R A^{(p)}) \underline{\otimes}_R (B^{(1)} \underline{\otimes}_R \cdots \underline{\otimes}_R B^{(q)}) \tag{3.4.8}$$

are isomorphic graded R-algebras under an isomorphism which matches $a^{(1)} \otimes \cdots \otimes a^{(p)} \otimes b^{(1)} \otimes \cdots \otimes b^{(q)}$ *with* $(a^{(1)} \otimes \cdots \otimes a^{(p)}) \otimes (b^{(1)} \otimes \cdots \otimes b^{(q)})$.

Proof. We already know that there is an isomorphism f, of graded algebras, which maps $A^{(1)} \otimes \cdots \otimes A^{(p)} \otimes B^{(1)} \otimes \cdots \otimes B^{(q)}$ onto $(A^{(1)} \otimes \cdots \otimes A^{(p)}) \otimes (B^{(1)} \otimes \cdots \otimes B^{(q)})$ and matches elements in the way described. It is therefore enough to show that f behaves properly with respect to modified products.

Suppose then that $I=(i_1, i_2, \ldots, i_p)$, $V=(v_1, v_2, \ldots, v_p)$, $J=(j_1, j_2, \ldots, j_q)$ and $W=(w_1, w_2, \ldots, w_q)$ are sequences of integers. We put $IJ=(i_1, \ldots, i_p, j_1, \ldots, j_q)$ and define VW similarly. We also put $|J|=j_1+j_2+\cdots+j_q$ and $|V|=v_1+v_2+\cdots+v_p$.

Let $x=a_{i_1} \otimes \cdots \otimes a_{i_p} \otimes b_{j_1} \otimes \cdots \otimes b_{j_q}$ and $y=\alpha_{v_1} \otimes \cdots \otimes \alpha_{v_p} \otimes \beta_{w_1} \otimes \cdots \otimes \beta_{w_q}$ belong to $A^{(1)}_{i_1} \otimes \cdots \otimes A^{(p)}_{i_p} \otimes B^{(1)}_{j_1} \otimes \cdots \otimes B^{(q)}_{j_q}$ and $A^{(1)}_{v_1} \otimes \cdots \otimes A^{(p)}_{v_p} \otimes B^{(1)}_{w_1} \otimes \cdots \otimes B^{(q)}_{w_q}$ respectively. The modified product of x and y is

$$\varepsilon(IJ, VW) a_{i_1}\alpha_{v_1} \otimes \cdots \otimes a_{i_p}\alpha_{v_p} \otimes b_{j_1}\beta_{w_1} \otimes \cdots \otimes b_{j_q}\beta_{w_q}$$

and the image of this under f is

$$\varepsilon(IJ, VW)(a_{i_1}\alpha_{v_1} \otimes \cdots \otimes a_{i_p}\alpha_{v_p}) \otimes (b_{j_1}\beta_{w_1} \otimes \cdots \otimes b_{j_q}\beta_{w_q}). \tag{3.4.9}$$

Thus the theorem will follow if we can show that (3.4.9) is the product of $(a_{i_1} \otimes \cdots \otimes a_{i_p}) \otimes (b_{j_1} \otimes \cdots \otimes b_{j_q})$ and $(\alpha_{v_i} \otimes \cdots \otimes \alpha_{v_p}) \otimes (\beta_{w_1} \otimes \cdots \otimes \beta_{w_q})$ in (3.4.8). However, this latter product is

$$(-1)^{|J||V|} \xi \otimes \eta,$$

where ξ is the product of $a_{i_1} \otimes \cdots \otimes a_{i_p}$ and $\alpha_{v_1} \otimes \cdots \otimes \alpha_{v_p}$ in $A^{(1)} \underline{\otimes} A^{(2)} \underline{\otimes} \cdots \underline{\otimes} A^{(p)}$, and η is the product of $b_{j_1} \otimes \cdots \otimes b_{j_q}$ and $\beta_{w_1} \otimes \cdots \otimes \beta_{w_q}$ in $B^{(1)} \underline{\otimes} B^{(2)} \underline{\otimes} \cdots \underline{\otimes} B^{(q)}$. Accordingly

$$(-1)^{|J||V|} \xi \otimes \eta$$

is just

$$(-1)^{|J||V|} \varepsilon(I, V) \varepsilon(J, W)(a_{i_1}\alpha_{v_1} \otimes \cdots \otimes a_{i_p}\alpha_{v_p}) \otimes (b_{j_1}\beta_{w_1} \otimes \cdots \otimes b_{j_q}\beta_{w_q})$$

and therefore the proof will be complete if we show that

$$\varepsilon(IJ, VW) = (-1)^{|J||V|} \varepsilon(I, V) \varepsilon(J, W).$$

However, this follows readily from (3.4.2) and (3.4.3).

We make one final observation concerning modified products. Suppose that

$$f_i: A^{(i)} \to B^{(i)} \quad (i=1,2,\dots,p)$$

are homomorphisms of graded algebras. Then of course $f_1 \otimes f_2 \otimes \cdots \otimes f_p$ is also a homomorphism

$$A^{(1)} \otimes A^{(2)} \otimes \cdots \otimes A^{(p)} \to B^{(1)} \otimes B^{(2)} \otimes \cdots \otimes B^{(p)}$$

of graded algebras. But modified and unmodified tensor products have the same R-module structure. Hence the same mapping $f_1 \otimes f_2 \otimes \cdots \otimes f_p$ is a homomorphism

$$A^{(1)} \underline{\otimes}_R A^{(2)} \underline{\otimes}_R \cdots \underline{\otimes}_R A^{(p)}$$
$$\to B^{(1)} \underline{\otimes}_R B^{(2)} \underline{\otimes}_R \cdots \underline{\otimes}_R B^{(p)} \tag{3.4.10}$$

of R-modules. This certainly preserves degrees and identity elements. *We claim that in fact* (3.4.10) *is a homomorphism of graded algebras.* To see this let x and y belong to $A^{(1)} \underline{\otimes} A^{(2)} \underline{\otimes} \cdots \underline{\otimes} A^{(p)}$. We have to show that the image of their modified product is the modified product of their images. For this we may assume that $x = a_{i_1} \otimes a_{i_2} \otimes \cdots \otimes a_{i_p}$ and $y = \alpha_{j_1} \otimes \alpha_{j_2} \otimes \cdots \otimes \alpha_{j_p}$ belong to $A_{i_1}^{(1)} \otimes \cdots \otimes A_{i_p}^{(p)}$ and $A_{j_1}^{(1)} \otimes \cdots \otimes A_{j_p}^{(p)}$ respectively, where we have made use of our usual notation. However, in this case the desired result follows from the formula for the modified product (see (3.4.5)).

3.5 Anticommutative algebras

Let $\{A_n\}_{n \in \mathbb{Z}}$ be a grading on an R-algebra A. We say that the graded algebra is *anticommutative* provided

$$\alpha_n a_m = (-1)^{mn} a_m \alpha_n \tag{3.5.1}$$

whenever a_m and α_n are homogeneous elements of degrees m and n respectively, and provided also that

$$a_m^2 = 0 \tag{3.5.2}$$

whenever the degree of a_m is odd. Thus we see that in an anticommutative algebra the square of a homogeneous element of degree one is zero. As the next lemma shows there is an important special situation where the converse holds.

Lemma 3. *Let the graded R-algebra A be generated (as an R-algebra) by its module A_1 of homogeneous elements of degree one. If now $a_1^2 = 0$ for every $a_1 \in A_1$, then A is anticommutative.*

Proof. From Lemma 1 we see that $A_n = 0$ for all $n < 0$ and also that A_0 is contained in the centre of A. Hence in verifying (3.5.1) and (3.5.2) we may assume that $m \geq 1$ and $n \geq 1$. Note that if $x, y \in A_1$, then from $(x+y)^2 = 0$,

$x^2=0$ and $y^2=0$ it follows that

$$xy+yx=0. \tag{3.5.3}$$

Now suppose that $u=x_1x_2\ldots x_m$, $v=y_1y_2\ldots y_n$, where every x_i and y_j belongs to A_1. Then, by (3.5.3),

$$\begin{aligned} vu &= y_1y_2\ldots y_nx_1x_2\ldots x_m \\ &= (-1)^{mn}x_1x_2\ldots x_my_1y_2\ldots y_n=(-1)^{mn}uv. \end{aligned}$$

But every element of A_m (respectively A_n) is a sum of elements such as u (respectively v) so the formula (3.5.1) holds quite generally.

From here on we suppose that $m\geq 1$ and is odd. Let a_m belong to A_m. Then $a_m=u_1+u_2+\cdots+u_s$, where each u_i is a product of m elements of A_1, and $u_i^2=0$ as is clear from (3.5.3) and the second hypothesis of the lemma. Again, by what has just been proved,

$$u_ju_i=(-1)^{m^2}u_iu_j=-u_iu_j$$

and now it follows that $a_m^2=0$. Thus the lemma is proved.

Theorem 7. *The modified tensor product of a finite number of anticommutative R-algebras is itself an anticommutative R-algebra.*

Proof. Let A and B be anticommutative R-algebras. If we can show that $A\,\underline{\otimes}\,B$ is anticommutative, then the general result will follow by virtue of Theorem 6.

Let $x=a_r\otimes b_s$, $y=\alpha_\rho\otimes\beta_\sigma$, where a_r, α_ρ are homogeneous elements of A and b_s, β_σ are homogeneous elements of B, and where the degree of each element is indicated by its suffix. If $x*y$ denotes the modified product of x and y, then

$$x*y=(-1)^{s\rho}a_r\alpha_\rho\otimes b_s\beta_\sigma=(-1)^{s\rho+r\rho+s\sigma}\alpha_\rho a_r\otimes\beta_\sigma b_s.$$

On the other hand

$$y*x=(-1)^{r\sigma}\alpha_\rho a_r\otimes\beta_\sigma b_s$$

so that

$$y*x=(-1)^{(r+s)(\rho+\sigma)}x*y.$$

But in $A\,\underline{\otimes}\,B$ the element x has degree $r+s=m$ say whereas y has degree $\rho+\sigma=n$ say. Consequently $y*x=(-1)^{mn}x*y$.

Now suppose that m is odd. Then one of r and s is odd and therefore the modified square of x namely

$$(-1)^{sr+r^2+s^2}a_r^2\otimes b_s^2$$

is zero.

Thus for $A\,\underline{\otimes}\,B$ we have verified the anticommutativity conditions for homogeneous elements of a special form. That these conditions hold for general homogeneous elements follows by linearity.

3.6 Covariant extension of an algebra

The process of covariant extension as applied to modules was explained in Section (2.5). Here we shall examine it in relation to algebras.

Let R, S be commutative rings and assume that we are given a ring-homomorphism

$$\omega: R \rightarrow S \tag{3.6.1}$$

that preserves identity elements. Assume further that A is an R-algebra. From Section (2.5) we know that $A \otimes_R S$ is an S-module. On the other hand S is an R-algebra (with ω as its structural homomorphism) so that $A \otimes_R S$ is an R-algebra and in particular it is a ring. If we examine the S-module structure and the ring structure on $A \otimes_R S$ we find that they interact in such a way that we have an S-algebra. Thus *the covariant extension of the R-algebra A is the S-algebra $A \otimes_R S$.* Note that the structural homomorphism $S \rightarrow A \otimes_R S$ maps s into $1_A \otimes s$.

Now suppose that $\{A_n\}_{n \in \mathbb{Z}}$ is a grading on A. By Theorem 5 of Chapter 2,

$$A \otimes_R S = \sum_{n \in \mathbb{Z}} (A_n \otimes_R S) \quad \text{(d.s.)}. \tag{3.6.2}$$

Of course according to the theorem just quoted the right-hand side is to be understood as a direct sum of R-modules. However, $A \otimes_R S$ and $A_n \otimes_R S$ are S-modules so that (3.6.2) is also a direct sum of S-modules. Moreover the product of x in $A_m \otimes_R S$ and y in $A_n \otimes_R S$ belongs to $A_{m+n} \otimes_R S$. Accordingly the S-algebra $A \otimes_R S$ is graded by $\{A_n \otimes_R S\}_{n \in \mathbb{Z}}$.

Let us turn our attention to the tensor product, over R, of R-algebras $A_1, A_2, \ldots, A_p$. (For the moment these algebras need not be graded.) By Theorem 9 of Chapter 2, there is an isomorphism

$$\begin{aligned}(A_1 \otimes_R S) \otimes_S (A_2 \otimes_R S) \otimes_S \cdots \otimes_S (A_p \otimes_R S)\\ \approx (A_1 \otimes_R A_2 \otimes_R \cdots \otimes_R A_p) \otimes_R S\end{aligned} \tag{3.6.3}$$

of S-modules in which $(a_1 \otimes s_1) \otimes (a_2 \otimes s_2) \otimes \cdots \otimes (a_p \otimes s_p)$ corresponds to $(a_1 \otimes a_2 \otimes \cdots \otimes a_p) \otimes s_1 s_2 \ldots s_p$. This isomorphism is compatible with multiplication and therefore (3.6.3) is actually an isomorphism of S-algebras. Thus, *for algebras, covariant extension commutes with the formation of tensor products.*

Finally suppose that each of the R-algebras $A_1, A_2, \ldots, A_p$ is graded. Then the two sides of (3.6.3) are graded S-algebras and it is easy to see that the isomorphism preserves degrees. Hence when *$A_1, A_2, \ldots, A_p$ are graded R-algebras* (3.6.3) *is an isomorphism of graded S-algebras.*

3.7 Derivations and skew derivations

Let $\{A_n\}_{n \in \mathbb{Z}}$ be a grading on an R-algebra A and let k be an integer.

A mapping $h: A \to A$ is said to be of *degree* k if it raises the degree of each homogeneous element by this amount.

By a *derivation* on A we shall understand an R-homomorphism $D: A \to A$, of degree -1, which satisfies

$$D(\alpha\beta) = (D\alpha)\beta + \alpha(D\beta) \tag{3.7.1}$$

for all α, β in A. Note that if D is a derivation, then

$$D(\alpha_1\alpha_2 \ldots \alpha_n) = \sum_{i=1}^{n} \alpha_1 \ldots \alpha_{i-1}(D\alpha_i)\alpha_{i+1} \ldots \alpha_n. \tag{3.7.2}$$

Suppose that D, D' are derivations on A and that $r \in R$. It is easily checked that the endomorphisms $D + D'$ and rD of A are themselves derivations. In fact the derivations on A form an R-module. Of course $D \circ D'$ is an endomorphism of A, but (usually) it is not a derivation.

Lemma 4. *Suppose that the graded R-algebra A is generated (as an R-algebra) by its homogeneous elements of degree one, and let D, D' be derivations on A. Then $D' \circ D = D \circ D'$. If D and D' agree on all homogeneous elements of degree one, then $D = D'$.*

Proof. Let $\alpha_1, \alpha_2, \ldots, \alpha_n$, where $n \geq 2$, belong to A_1. Then $D\alpha_i$ and $D'\alpha_i$ belong to A_0 and are therefore in the centre of A. Also $(D' \circ D)\alpha_i = 0$ because $A_s = 0$ for all $s < 0$. Next, from (3.7.2) we have

$$\begin{aligned}(D' \circ D)&(\alpha_1\alpha_2 \ldots \alpha_n)\\ &= \sum_{i<j} (D(\alpha_i)D'(\alpha_j) + D(\alpha_j)D'(\alpha_i))\alpha_1 \ldots \hat{\alpha}_i \ldots \hat{\alpha}_j \ldots \alpha_n,\end{aligned}$$

where the symbol $\hat{}$ indicates that the term over which it is placed is to be omitted. It follows that

$$(D' \circ D)(\alpha_1\alpha_2 \ldots \alpha_n) = (D \circ D')(\alpha_1\alpha_2 \ldots \alpha_n)$$

so that $D' \circ D = D \circ D'$ as stated. The final assertion of the lemma also follows from (3.7.2).

Derivations play an important role in the theory of commutative algebras. For anticommutative algebras the concept needs to be modified.

By a *skew derivation* on A we shall mean an R-homomorphism $\Delta: A \to A$, of degree -1, which satisfies

$$\Delta(\alpha\beta) = (\Delta\alpha)\beta + (-1)^m\alpha(\Delta\beta) \tag{3.7.3}$$

for all $\alpha \in A_m$ and $\beta \in A$. The skew derivations on A also form an R-module.

Lemma 5. *Let Δ be a skew derivation on the graded R-algebra A and let $\alpha_1, \alpha_2, \ldots, \alpha_n$ be homogeneous elements of degree one. Then*

$$\Delta(\alpha_1\alpha_2 \ldots \alpha_n) = \sum_{i=1}^{n} (-1)^{i+1}\alpha_1 \ldots (\Delta\alpha_i) \ldots \alpha_n. \tag{3.7.4}$$

Proof. Since

$$\Delta(\alpha_1\alpha_2 \dots \alpha_n) = (\Delta(\alpha_1\alpha_2 \dots \alpha_{n-1}))\alpha_n + (-1)^{n-1}\alpha_1\alpha_2 \dots \alpha_{n-1}(\Delta\alpha_n)$$

the lemma follows by induction.

Lemma 6. *Let the graded R-algebra A be generated (as an R-algebra) by its homogeneous elements of degree one, and let Δ, Δ' be skew derivations on A. Then $\Delta \circ \Delta = 0$ and $\Delta' \circ \Delta + \Delta \circ \Delta' = 0$. If Δ and Δ' coincide on the module of homogeneous elements of degree one, then $\Delta = \Delta'$.*

Proof. Let $\alpha_1, \alpha_2, \dots, \alpha_n$ be homogeneous elements of degree one. Then $\Delta\alpha_i$ belongs to the centre of A. Two applications of (3.7.4) now show that $(\Delta \circ \Delta)(\alpha_1\alpha_2 \dots \alpha_n) = 0$ whence $\Delta \circ \Delta = 0$. This proves the first assertion and the second follows by expanding $(\Delta + \Delta') \circ (\Delta + \Delta') = 0$. The final assertion is clear from (3.7.4).

Derivations and skew derivations both behave well in relation to the process of covariant extension. To be precise let $\omega: R \to S$ be a homomorphism of commutative rings and let D (respectively Δ) be a derivation (respectively skew derivation) on a graded R-algebra A. Then $D \otimes S$ (respectively $\Delta \otimes S$) is a derivation (respectively skew derivation) on the graded S-algebra $A \otimes_R S$.

3.8 Comments and exercises

Perhaps the first comment to be made concerning the notion of an algebra is that it subsumes more than appears at first sight. Thus if Ω is any ring with an identity element, then Ω is an *algebra* over the integers. Here the structural homomorphism $\mathbb{Z} \to \Omega$ maps the integer n into $n1_\Omega$. Additional comments, supplemented by exercises, are to be found below, and, as in Chapters 1 and 2, the exercises for which solutions are provided are marked with an asterisk.

We begin by noting that if $X_1, X_2, \dots, X_n$ are indeterminates, then the polynomial ring $R[X_1, X_2, \dots, X_n]$ is a *commutative* R-algebra.

Exercise 1. *Show that an R-algebra A can be generated (as an algebra) by a single element if and only if it is a homomorphic image of the algebra $R[X]$, where X is an indeterminate.*

Observe that this exercise shows that an algebra which can be generated by a single element is necessarily commutative.

Now let Σ be a multiplicatively closed subset of R. We explained in Section (2.6) how the ring $\Sigma^{-1}(R)$ of fractions is constructed and we noted that there is a ring-homomorphism $R \to \Sigma^{-1}(R)$ in which r of R is mapped into the fraction $r/1$. As a result of this homomorphism $\Sigma^{-1}(R)$ becomes a commutative R-algebra.

This said, let Σ_1 and Σ_2 be two multiplicatively closed subsets of R. Then $\Sigma_1^{-1}(R)$ and $\Sigma_2^{-1}(R)$ are R-algebras so we can form their tensor product

$$\Sigma_1^{-1}(R) \otimes_R \Sigma_2^{-1}(R) \tag{3.8.1}$$

and this too will be a commutative R-algebra. On the other hand, the set of all products $\sigma_1\sigma_2$, where $\sigma_1 \in \Sigma_1$ and $\sigma_2 \in \Sigma_2$, is again a multiplicatively closed subset of R. Let this new multiplicatively closed subset be denoted by $\Sigma_1\Sigma_2$ and let us form the R-algebra

$$(\Sigma_1\Sigma_2)^{-1}(R). \tag{3.8.2}$$

The next exercise shows that (3.8.1) and (3.8.2) are virtually identical.

Exercise 2*. *Let Σ_1 and Σ_2 be multiplicatively closed subsets of R and put $\Sigma = \Sigma_1\Sigma_2$. Establish an isomorphism*

$$\Sigma_1^{-1}(R) \otimes_R \Sigma_2^{-1}(R) \approx \Sigma^{-1}(R)$$

of R-algebras.

The existence of a grading on an R-algebra has some interesting consequences. Let A be a graded R-algebra and let $\{A_n\}_{n\in\mathbb{Z}}$ be its grading. First we make the rather trivial observation that *if α belongs to the centre of A, then all its homogeneous components belong to the centre as well.* To see this let

$$\alpha = \cdots + \alpha_{-1} + \alpha_0 + \alpha_1 + \alpha_2 + \cdots$$

be the decomposition of α into its homogeneous components, where the suffix indicates the degree of an individual component. If now $a_m \in A_m$, then from $\alpha a_m = a_m \alpha$ we obtain $\alpha_n a_m = a_m \alpha_n$ by comparing components of degree $m+n$. But now it follows that $\alpha_n x = x\alpha_n$ for all $x \in A$, i.e. α_n belongs to the centre of A.

Exercise 3*. *Let A be a non-negatively graded R-algebra and let α belong to the centre of A. If $\alpha^2 = \alpha$ show that α is homogeneous of degree zero.*

Of great importance is the notion of the *free algebra generated by a given set*. To explain what this is let X be an arbitrary non-empty set. We consider formal products $x_1x_2 \ldots x_n$ $(n \geq 0)$ of elements of X or, to use an alternative terminology, *words* formed by using the elements of X as letters. (Note that, because the possibility $n=0$ is allowed, we have included the *empty* word.) Next we form the free R-module which has the set of words as a base, and afterwards turn the free module into an R-algebra by defining the product of two of its elements by means of the self-explanatory formula

$$(\sum rx_1x_2 \ldots x_n)(\sum r'x_1'x_2' \ldots x_q') = \sum\sum (rr')x_1 \ldots x_nx_1' \ldots x_q'.$$

The resulting algebra is the free R-algebra generated by X. (Naturally the free algebra generated by the *empty* set is taken to be R itself.)

Let A be this algebra. For $n \geq 0$ let A_n be the R-submodule generated by all words with n letters, and put $A_q = 0$ when $q < 0$. It is easy to check that $\{A_s\}_{s \in \mathbb{Z}}$ is a non-negative grading on A. Note that $A_0 = R$, that A_1 is the free R-module generated by X, and that this module generates the algebra. In the next chapter where with each module we shall associate an algebra called its tensor algebra, we shall see that A is the tensor algebra of the free module generated by X.

Another interesting example arises in the following way. Let M be an R-module and this time consider the set A formed by all 2×2 'matrices'

$$\begin{bmatrix} r & m \\ 0 & r \end{bmatrix}$$

where $r \in R$ and $m \in M$. If addition and multiplication on A are now defined by

$$\begin{bmatrix} r & m \\ 0 & r \end{bmatrix} + \begin{bmatrix} r' & m' \\ 0 & r' \end{bmatrix} = \begin{bmatrix} r+r' & m+m' \\ 0 & r+r' \end{bmatrix}$$

and

$$\begin{bmatrix} r & m \\ 0 & r \end{bmatrix} \begin{bmatrix} r' & m' \\ 0 & r' \end{bmatrix} = \begin{bmatrix} rr' & rm'+r'm \\ 0 & rr' \end{bmatrix}$$

respectively, then it is easily checked that A is a commutative ring with

$$\begin{bmatrix} 1 & 0 \\ 0 & 1 \end{bmatrix}$$

as its identity element. Furthermore, if for $\rho \in R$ we put

$$\rho \begin{bmatrix} r & m \\ 0 & r \end{bmatrix} = \begin{bmatrix} \rho r & \rho m \\ 0 & \rho r \end{bmatrix},$$

then A is turned into a commutative R-algebra.

Now let A_0 consist of all matrices of the form

$$\begin{bmatrix} r & 0 \\ 0 & r \end{bmatrix},$$

let A_1 consist of all matrices

$$\begin{bmatrix} 0 & m \\ 0 & 0 \end{bmatrix},$$

and for $n \neq 0, 1$ put $A_n = 0$. Then $\{A_q\}_{q \in \mathbb{Z}}$ is a family of R-submodules of A.

Exercise 4. *With the above notation, show that $\{A_q\}_{q \in \mathbb{Z}}$ is an algebra-grading for the R-algebra formed by all matrices of the form*

$$\begin{bmatrix} r & m \\ 0 & r \end{bmatrix},$$

where $r \in R$ and $m \in M$.

Let $A^{(1)}, A^{(2)}, \ldots, A^{(p)}$ be graded R-algebras. We have discussed two graded algebras that can be obtained from them by using tensor products. These algebras are the ordinary tensor product $A^{(1)} \otimes A^{(2)} \otimes \cdots \otimes A^{(p)}$ and the modified tensor product $A^{(1)} \underline{\otimes} A^{(2)} \underline{\otimes} \cdots \underline{\otimes} A^{(p)}$. The latter will also be known as the *twisted* tensor product.

In establishing (3.3.13) we showed that the operation of combining graded algebras by means of the ordinary tensor product is *associative* and the discussion of (3.3.14) shows that it is commutative as well. Furthermore, from (3.3.15) and (3.3.16), we see that when R is endowed with the trivial grading it acts like an identity in this context.

For modified, that is twisted, tensor products the facts are not so transparent. We have indeed shown (Theorem 6) that the associative law continues to be satisfied, but so far we have not examined whether modified tensor multiplication is commutative. This will be our next concern.

Let A and B be graded R-algebras and $\{A_n\}_{n\in\mathbb{Z}}$ and $\{B_n\}_{n\in\mathbb{Z}}$ their respective gradings. There is an isomorphism $A_m \otimes B_n \approx B_n \otimes A_m$ in which $a_m \otimes b_n$ is matched with $b_n \otimes a_m$. (Here, of course, $a_m \in A_m$ and $b_n \in B_n$.) It follows that there is an isomorphism

$$T_{mn}: A_m \otimes B_n \xrightarrow{\sim} B_n \otimes A_m \tag{3.8.3}$$

in which

$$T_{mn}(a_m \otimes b_n) = (-1)^{mn} b_n \otimes a_m. \tag{3.8.4}$$

But $A \otimes B = \sum A_m \otimes B_n$ (d.s.) and $B \otimes A = \sum B_n \otimes A_m$ (d.s.) so we can combine the various T_{mn} to obtain an isomorphism

$$T: A \otimes B \xrightarrow{\sim} B \otimes A \tag{3.8.5}$$

of R-modules which is such that

$$T(a_m \otimes b_n) = (-1)^{mn} b_n \otimes a_m \tag{3.8.6}$$

whenever $a_m \in A_m$ and $b_n \in B_n$. T is known as the *twisting* isomorphism. Now $A \otimes B$ and $A \underline{\otimes} B$ are identical as R-modules and the same is true of $B \otimes A$ and $B \underline{\otimes} A$. Consequently T provides a bijection

$$T: A \underline{\otimes} B \to B \underline{\otimes} A. \tag{3.8.7}$$

Exercise 5*. *Let A and B be graded R-algebras. Show that the twisting isomorphism $T: A \otimes B \xrightarrow{\sim} B \otimes A$ produces an isomorphism*

$$T: A \underline{\otimes} B \xrightarrow{\sim} B \underline{\otimes} A$$

of graded R-algebras.

The next exercise extends the result embodied in Exercise 5 to twisted tensor products with an arbitrary number of factors.

Exercise 6*. *Let $A^{(1)}, A^{(2)}, \ldots, A^{(p)}$ be graded R-algebras and let $(s_1, s_2, \ldots, s_p)$ be a permutation of $(1, 2, \ldots, p)$. Show that*

$$A^{(1)} \underline{\otimes} A^{(2)} \underline{\otimes} \cdots \underline{\otimes} A^{(p)}$$

and

$$A^{(s_1)} \underline{\otimes} A^{(s_2)} \underline{\otimes} \cdots \underline{\otimes} A^{(s_p)}$$

are isomorphic graded algebras.

Let A be an R-algebra. The mapping $A \times A \to A$ in which (a, α) is taken into $a\alpha$ is certainly bilinear and so it induces an R-linear mapping $\mu: A \otimes A \to A$ in which $\mu(a \otimes \alpha) = a\alpha$. If A is a commutative R-algebra, then $\mu: A \otimes A \to A$ is a homomorphism of algebras, and if A is a commutative graded algebra, then μ is a homomorphism of graded algebras. The next exercise deals with the corresponding result for anticommutative algebras. Note that μ can also be regarded as an R-linear mapping of $A \underline{\otimes} A$ into A.

Exercise 7*. *Let A be an anticommutative R-algebra. Show that the mapping $\mu: A \underline{\otimes} A \to A$ in which $\mu(a \otimes \alpha) = a\alpha$ is a homomorphism of graded algebras.*

Ordinary and modified tensor products of graded algebras have so many features in common that the idea of finding an approach that will deal with both types simultaneously has obvious attractions, and in fact something can be done on these lines as will now be explained.

Let A and B be graded R-algebras with $\{A_n\}_{n \in \mathbb{Z}}$ and $\{B_n\}_{n \in \mathbb{Z}}$ their respective gradings. Suppose that $A_n = 0$ whenever n is an *odd* integer. Then $A \otimes B$ and $A \underline{\otimes} B$ not only coincide as modules, but ordinary and modified multiplication are the same. Consequently $A \otimes B$ and $A \underline{\otimes} B$ are identical algebras and we may write

$$A \otimes B = A \underline{\otimes} B. \tag{3.8.8}$$

Likewise (with the same assumption on A) we have

$$B \otimes A = B \underline{\otimes} A. \tag{3.8.9}$$

For example, (3.8.8) and (3.8.9) hold if the grading on A is trivial. In particular, if R is endowed with the trivial grading, then, for any graded R-algebra B, $R \underline{\otimes} B = R \otimes B$ and $B \underline{\otimes} R = B \otimes R$ and therefore we have isomorphisms

$$R \underline{\otimes} B \approx B \tag{3.8.10}$$

and

$$B \underline{\otimes} R \approx B \tag{3.8.11}$$

of graded algebras. These isomorphisms are effected in the usual way so that, for instance, in (3.8.10) $r \otimes b$, where $r \in R$ and $b \in B$, corresponds to rb.

Suppose now that C is a graded R-algebra and let C_n denote its submodule of homogeneous elements of degree n. Put

$$C'_n = \begin{cases} 0 & \text{if } n \text{ is odd,} \\ C_{n/2} & \text{if } n \text{ is even.} \end{cases}$$

Then $C = \sum C'_n$ (d.s.) and indeed $\{C'_n\}_{n \in \mathbb{Z}}$ is an algebra-grading on C. Accordingly we have a new graded R-algebra, C' say. The connection between C' and C is described by saying that C' is obtained from C by *doubling degrees.*

Exercise 8. *Let C and D be graded R-algebras. Show that*

$$C' \otimes D' = (C \otimes D)',$$

where C', D' and $(C \otimes D)'$ are obtained from C, D and $C \otimes D$ respectively by doubling degrees.

Let us retain, for the moment, the notation of Exercise 8. Then, by (3.8.8), $C' \otimes D' = C' \underline{\otimes} D'$ and therefore

$$C' \underline{\otimes} D' = (C \otimes D)'. \tag{3.8.12}$$

Now the operation of doubling degrees is easily reversed. Consequently (3.8.12) provides a means whereby an ordinary tensor product can be expressed in terms of a modified tensor product and this, on occasions, offers some advantages.

We next observe that the notions of *derivation* and *skew derivation*, as defined in Section (3.7), can be considerably extended. For instance, let A be a graded R-algebra. Then by a *generalized derivation* of degree i we understand an R-linear mapping $D: A \to A$ of degree i such that

$$D(a\alpha) = (Da)\alpha + a(D\alpha)$$

for all a, α in A. Thus the derivations considered in the main text are precisely the generalized derivations of degree -1.

Exercise 9. *Let D and D' be generalized derivations, on the graded R-algebra A, of degrees i and j respectively. Show that $D' \circ D - D \circ D'$ is a generalized derivation of degree $i+j$.*

As is to be expected, generalizing the notion of a skew derivation is a more complicated matter. It is helpful to make some preparatory observations.

Once again let $\{A_n\}_{n \in \mathbb{Z}}$ be a grading on an R-algebra A. For each $n \in \mathbb{Z}$ we can define an automorphism J_n, of A_n, by putting $J_n(x) = (-1)^n x$ for all $x \in A_n$; and then these automorphisms can be combined to give an automorphism

$$J: A \xrightarrow{\sim} A \tag{3.8.13}$$

of the R-module A which satisfies

$$J(a_n) = (-1)^n a_n \tag{3.8.14}$$

for all a_n in A_n. Actually J is more than a module-automorphism as the next exercise shows.

Exercise 10. *Show that J, as defined in* (3.8.13), *is an automorphism of the R-algebra A.*

Indeed this is not the end of the matter. Not only is J an automorphism of the algebra, but J^2 is the identity automorphism, so that, in other words, J is an *involution* of the graded algebra. It is known as the *main involution* of A. Note that if $i \in \mathbb{Z}$, then J^i is the identity automorphism if i is even whereas it is J itself if i is odd. Note too that if $a_m \in A_m$, then

$$J^i(a_m) = (-1)^{im} a_m. \tag{3.8.15}$$

After these preliminaries let $\Delta: A \to A$ be an R-linear mapping of degree -1. Then Δ is a skew derivation (as defined in Section (3.7)) if and only if

$$\Delta(a\alpha) = (\Delta a)\alpha + J(a)\Delta(\alpha) \tag{3.8.16}$$

for all a, α in A.

We are now ready to generalize the notion of a skew derivation. Specifically a mapping $\Delta: A \to A$ will be called a *generalized skew derivation* of degree i if it is R-linear of degree i and

$$\Delta(a\alpha) = (\Delta a)\alpha + J^i(a)\Delta(\alpha) \tag{3.8.17}$$

for a, α in A. Consequently the skew derivations which occur in the main text are none other than the generalized skew derivations of degree -1.

Exercise 11*. *Let Δ and Δ' be generalized skew derivations, of degrees i and j respectively, on the graded R-algebra A. Show that*

$$\Delta' \circ \Delta - (-1)^{ij} \Delta \circ \Delta'$$

is a generalized skew derivation of degree $i+j$. Show also that if the integer i is odd, then $\Delta \circ \Delta$ is a generalized skew derivation of degree $2i$.

One last remark. We saw earlier that the device of doubling degrees enables us to express the ordinary tensor product of two graded algebras in terms of a modified tensor product (see (3.8.12)). To this extent it is not necessary to develop separate theories for the two sorts of product. In a similar way the theory of generalized derivations can be included in the theory of generalized skew derivations. For suppose that A is a graded R-algebra and let A' be the algebra obtained from A by doubling the degrees of its homogeneous elements. Now suppose that D is an R-linear mapping of A into itself. Then D is a generalized derivation on A of degree i if and only if, considered as an endomorphism of A', it is a generalized skew derivation of degree $2i$.

3.9 Solutions to selected exercises

Exercise 2. *Let Σ_1 and Σ_2 be multiplicatively closed subsets of R and put $\Sigma = \Sigma_1\Sigma_2$. Establish an isomorphism*

$$\Sigma_1^{-1}(R) \otimes_R \Sigma_2^{-1}(R) \approx \Sigma^{-1}(R)$$

of R-algebras.

Solution. In what follows σ_1, σ_1' denote elements of Σ_1; σ_2, σ_2' elements of Σ_2; and r, r', ρ, ρ' elements of R.

If $r/\sigma_1 = r'/\sigma_1'$ in $\Sigma_1^{-1}(R)$ and $\rho/\sigma_2 = \rho'/\sigma_2'$ in $\Sigma_2^{-1}(R)$, then it is easily verified that

$$\frac{r\rho}{\sigma_1\sigma_2} = \frac{r'\rho'}{\sigma_1'\sigma_2'}$$

in $\Sigma^{-1}(R)$. Hence there is a well-defined mapping

$$\Sigma_1^{-1}(R) \times \Sigma_2^{-1}(R) \to \Sigma^{-1}(R)$$

in which $(r/\sigma_1, \rho/\sigma_2)$ is mapped into $r\rho/\sigma_1\sigma_2$, and a straightforward check shows that the mapping is bilinear. Accordingly there is an R-linear mapping

$$\phi\colon \Sigma_1^{-1}(R) \otimes_R \Sigma_2^{-1}(R) \to \Sigma^{-1}(R),$$

where $\phi(r/\sigma_1 \otimes \rho/\sigma_2) = r\rho/\sigma_1\sigma_2$.

We claim that ϕ is an isomorphism of R-algebras. Indeed it is obvious that ϕ is surjective and that it preserves identity elements. Next, if M is any R-module, then we know that each element of $M \otimes_R \Sigma_2^{-1}(R)$ can be written in the form $m \otimes (1/\sigma_2)$, where $m \in M$. Consequently every element of $\Sigma_1^{-1}(R) \otimes_R \Sigma_2^{-1}(R)$ can be expressed in the form $(r/\sigma_1) \otimes (\rho/\sigma_2)$. Let $x = (r/\sigma_1) \otimes (\rho/\sigma_2)$ and $x' = (r'/\sigma_1') \otimes (\rho'/\sigma_2')$. Then

$$\begin{aligned}\phi(xx') &= \phi\{(rr'/\sigma_1\sigma_1') \otimes (\rho\rho'/\sigma_2\sigma_2')\}\\ &= rr'\rho\rho'/\sigma_1\sigma_1'\sigma_2\sigma_2'\\ &= (r\rho/\sigma_1\sigma_2)(r'\rho'/\sigma_1'\sigma_2')\\ &= \phi(x)\phi(x')\end{aligned}$$

so that ϕ is a homomorphism of R-algebras.

Finally let $x = (r/\sigma_1) \otimes (\rho/\sigma_2)$ and suppose that $\phi(x) = 0$. Then $r\rho/\sigma_1\sigma_2 = 0$ in $\Sigma^{-1}(R)$ and therefore there exist $\sigma_1' \in \Sigma_1$ and $\sigma_2' \in \Sigma_2$ such that $\sigma_1'\sigma_2' r\rho = 0$. But now

$$\begin{aligned}x &= (\sigma_1' r/\sigma_1\sigma_1') \otimes (\sigma_2'\rho/\sigma_2\sigma_2')\\ &= (\sigma_1'\sigma_2' r\rho/\sigma_1\sigma_1') \otimes (1/\sigma_2\sigma_2')\\ &= 0.\end{aligned}$$

Thus ϕ is an injection and the verification that ϕ is an isomorphism is complete.

Exercise 3. *Let A be a non-negatively graded R-algebra and let α belong to the centre of A. If $\alpha^2 = \alpha$ show that α is homogeneous of degree zero.*

Solution. Let $\alpha = \alpha_0 + \alpha_1 + \alpha_2 + \cdots$, where α_m is homogeneous of degree m, and put $\beta = \alpha_1 + \alpha_2 + \cdots$ so that $\alpha = \alpha_0 + \beta$. Then $\alpha_0^2 = \alpha_0$ and both α_0 and β are in the centre of A. Now

$$(1-\alpha_0)\beta = (1-\alpha_0)(\alpha-\alpha_0) = (1-\alpha_0)\alpha$$

and from this we see that $(1-\alpha_0)\beta$ is equal to its own square. Accordingly $((1-\alpha_0)\beta)^m = (1-\alpha_0)\beta$ for all $m \geq 1$. But $(1-\alpha_0)\beta$ is a sum of homogeneous elements of strictly positive degrees and therefore we may conclude that $(1-\alpha_0)\beta = 0$, i.e. that $\beta = \alpha_0\beta$. Next

$$\alpha_0 + \beta = \alpha = \alpha^2 = (\alpha_0 + \beta)^2 = \alpha_0^2 + 2\alpha_0\beta + \beta^2$$

whence $\beta^2 = -\beta$. It follows that $\beta^m = (-1)^{m+1}\beta$ for all $m \geq 1$. However, this implies that $\beta = 0$ because β is a sum of homogeneous elements of strictly positive degrees. Thus $\alpha = \alpha_0$ and the solution is complete.

Exercise 5. *Let A and B be graded R-algebras. Show that the twisting isomorphism $T: A \otimes B \xrightarrow{\sim} B \otimes A$ produces an isomorphism*

$$T: A \underline{\otimes} B \xrightarrow{\sim} B \underline{\otimes} A$$

of graded algebras.

Solution. Let x and y belong to $A \underline{\otimes} B$. Since T preserves degrees it will be enough to show that $T(x * y) = T(x) * T(y)$, where an asterisk has been used to indicate a modified (as distinct from an ordinary) product. But T is an isomorphism of R-modules. We may therefore confine our attention to the case where $x = a_r \otimes b_s$ and $y = \alpha_\rho \otimes \beta_\sigma$; here a_r, α_ρ denote homogeneous elements of A and b_s, β_σ homogeneous elements of B, the degrees of these elements being indicated by their suffixes. On this understanding $x * y = (-1)^{s\rho} a_r\alpha_\rho \otimes b_s\beta_\sigma$ and therefore

$$\begin{aligned} T(x * y) &= (-1)^{s\rho}(-1)^{(r+\rho)(s+\sigma)} b_s\beta_\sigma \otimes a_r\alpha_\rho \\ &= (-1)^{rs + r\sigma + \rho\sigma} b_s\beta_\sigma \otimes a_r\alpha_\rho. \end{aligned}$$

On the other hand $T(x) = (-1)^{rs} b_s \otimes a_r$ and $T(y) = (-1)^{\rho\sigma}\beta_\sigma \otimes \alpha_\rho$. Consequently

$$\begin{aligned} T(x) * T(y) &= (-1)^{rs + \rho\sigma + r\sigma} b_s\beta_\sigma \otimes a_r\alpha_\rho \\ &= T(x * y) \end{aligned}$$

which is what we were seeking to prove.

Exercise 6. *Let $A^{(1)}, A^{(2)}, \ldots, A^{(p)}$ be graded R-algebras and let $(s_1, s_2, \ldots, s_p)$ be a permutation of $(1, 2, \ldots, p)$. Show that*

$$A^{(1)} \underline{\otimes} A^{(2)} \underline{\otimes} \cdots \underline{\otimes} A^{(p)} \tag{3.9.1}$$

and

$$A^{(s_1)} \underline{\otimes} A^{(s_2)} \underline{\otimes} \cdots \underline{\otimes} A^{(s_p)} \tag{3.9.2}$$

are isomorphic graded algebras.

Solution. Suppose that $1 \leq i < p$. By repeated applications of Theorem 6 together with the remarks made concerning (3.4.10), we obtain the following isomorphisms of graded algebras:

$$\begin{aligned} &A^{(1)} \underline{\otimes} \cdots \underline{\otimes} A^{(i-1)} \underline{\otimes} A^{(i)} \underline{\otimes} A^{(i+1)} \underline{\otimes} A^{(i+2)} \underline{\otimes} \cdots \underline{\otimes} A^{(p)} \\ &\approx (A^{(1)} \underline{\otimes} \cdots \underline{\otimes} A^{(i)} \underline{\otimes} A^{(i+1)}) \underline{\otimes} (A^{(i+2)} \underline{\otimes} \cdots \underline{\otimes} A^{(p)}) \\ &\approx \{(A^{(1)} \underline{\otimes} \cdots \underline{\otimes} A^{(i-1)}) \underline{\otimes} (A^{(i)} \underline{\otimes} A^{(i+1)})\} \underline{\otimes} (A^{(i+2)} \underline{\otimes} \cdots \underline{\otimes} A^{(p)}) \\ &\approx (A^{(1)} \underline{\otimes} \cdots \underline{\otimes} A^{(i-1)}) \underline{\otimes} (A^{(i)} \underline{\otimes} A^{(i+1)}) \underline{\otimes} (A^{(i+2)} \underline{\otimes} \cdots \underline{\otimes} A^{(p)}). \end{aligned}$$

In a similar manner we obtain an isomorphism

$$\begin{aligned} &A^{(1)} \underline{\otimes} \cdots \underline{\otimes} A^{(i-1)} \underline{\otimes} A^{(i+1)} \underline{\otimes} A^{(i)} \underline{\otimes} A^{(i+2)} \underline{\otimes} \cdots \underline{\otimes} A^{(p)} \\ &\quad \approx (A^{(1)} \underline{\otimes} \cdots \underline{\otimes} A^{(i-1)}) \underline{\otimes} (A^{(i+1)} \underline{\otimes} A^{(i)}) \underline{\otimes} (A^{(i+2)} \underline{\otimes} \cdots \underline{\otimes} A^{(p)}). \end{aligned}$$

But we know from Exercise 5 that $A^{(i)} \underline{\otimes} A^{(i+1)}$ and $A^{(i+1)} \underline{\otimes} A^{(i)}$ are isomorphic graded algebras. It follows that the algebra (3.9.1) remains unchanged up to an isomorphism of graded algebras if we interchange the *adjacent* factors $A^{(i)}$ and $A^{(i+1)}$. However, the factors may be brought into whatever order we please by a succession of adjacent interchanges and with this observation the solution is complete.

The reader may like to refine this result by constructing an *explicit* isomorphism between the algebras (3.9.1) and (3.9.2).

Exercise 7. *Let A be an anticommutative R-algebra. Show that the mapping $\mu: A \underline{\otimes} A \to A$ in which $\mu(a \otimes \alpha) = a\alpha$ is a homomorphism of graded algebras.*

Solution. It is clear that μ is R-linear and that it preserves identities and degrees. Let x and x' belong to $A \underline{\otimes} A$ and let $x * x'$ denote their product in $A \underline{\otimes} A$. We need only prove that $\mu(x * x') = \mu(x)\mu(x')$ and in doing this we may suppose that $x = a_r \otimes \alpha_s$, $x' = a'_\rho \otimes \alpha'_\sigma$, where $a_r, a'_\rho, \alpha_s, \alpha'_\sigma$ are homogeneous elements of A whose degrees are indicated by their suffixes. Now $x * x' = (-1)^{s\rho} a_r a'_\rho \otimes \alpha_s \alpha'_\sigma$ and therefore

$$\mu(x * x') = (-1)^{s\rho} a_r a'_\rho \alpha_s \alpha'_\sigma = a_r \alpha_s a'_\rho \alpha'_\sigma$$

because A is anticommutative. Accordingly $\mu(x * x') = \mu(x)\mu(x')$ as required.

Exercise 11. *Let Δ and Δ' be skew derivations, of degrees i and j respectively, on the graded R-algebra A. Show that*

$$\Delta' \circ \Delta - (-1)^{ij} \Delta \circ \Delta'$$

is a generalized skew derivation of degree $i+j$. Show also that if the integer i is odd, then $\Delta \circ \Delta$ is a generalized skew derivation of degree $2i$.

Solution. It is clear that $\Delta' \circ \Delta$ and $\Delta \circ \Delta'$ are endomorphisms of degree $i+j$ and therefore the same is true of $\Delta' \circ \Delta - (-1)^{ij}\Delta \circ \Delta'$. Now let a, α belong to A. Then $\Delta(a\alpha) = (\Delta a)\alpha + J^i(a)(\Delta\alpha)$ and therefore

$$\Delta'(\Delta(a\alpha)) = (\Delta'(\Delta a))\alpha + J^j(\Delta a)(\Delta'\alpha) + \Delta'(J^i(a))(\Delta\alpha) + J^j(J^i(a))(\Delta'(\Delta\alpha)).$$

Similarly

$$\Delta(\Delta'(a\alpha)) = (\Delta(\Delta' a))\alpha + J^i(\Delta' a)(\Delta\alpha) + \Delta(J^j(a))(\Delta'\alpha) + J^i(J^j(a))(\Delta(\Delta'\alpha)).$$

But $J^j(J^i(a)) = J^i(J^j(a)) = J^{i+j}(a)$ and therefore it will follow that $\Delta' \circ \Delta - (-1)^{ij}\Delta \circ \Delta'$ is a skew derivation of degree $i+j$ provided we show that

$$J^j(\Delta a) = (-1)^{ij}\Delta(J^j(a)) \tag{3.9.3}$$

and

$$J^i(\Delta' a) = (-1)^{ij}\Delta'(J^i(a)). \qquad (3.9.4))$$

Indeed it is enough to prove the first of these relations for then the second will follow by symmetry; and furthermore it will suffice to prove (3.9.3) when a is a *homogeneous* element.

Suppose therefore that $a \in A_m$. Then

$$\begin{aligned} J^j(\Delta a) &= (-1)^{j(m+i)}\Delta(a) \\ &= (-1)^{ij}\Delta((-1)^{jm}a) \\ &= (-1)^{ij}\Delta(J^j(a)) \end{aligned}$$

which is what we were aiming to prove.

From here on we shall suppose that i is an *odd* integer and that a, α are general elements of A. Then

$$\Delta(\Delta(a\alpha)) = (\Delta(\Delta a))\alpha + J^i(\Delta a)(\Delta\alpha) + \Delta(J^i(a))(\Delta\alpha) + J^{2i}(a)(\Delta(\Delta\alpha))$$

and so we shall have proved that $\Delta \circ \Delta$ is a skew derivation of degree $2i$ if we show that

$$J^i(\Delta a) + \Delta(J^i(a)) = 0.$$

This, however, is an immediate consequence of (3.9.3) because the integer i is odd.

4

The tensor algebra of a module

General remarks

In Section (1.3) we defined the tensor powers of an R-module M. In this chapter we shall show how the different tensor powers can be fitted together to produce an R-algebra. This algebra, which is known as the *tensor algebra* of M, contains M as a submodule. In fact the tensor algebra solves a certain universal problem concerned with algebras that contain a homomorphic image of M. The tensor algebra is particularly important in our context because it has, as homomorphic images, the exterior and symmetric algebras which we shall study later.

As in previous chapters R and S always denote *commutative* rings. All algebras are understood to be *associative*, rings and algebras are assumed to have identity elements, and homomorphisms of rings and algebras are required to preserve identity elements. Finally, we shall often omit the suffix from the tensor symbol $\otimes$ when the underlying ring can easily be inferred from the context.

4.1 The tensor algebra

Let M be an R-module. We shall define its tensor algebra in terms of a certain universal problem. To this end suppose that A is an R-algebra and that $\phi: M \rightarrow A$ is a homomorphism of R-modules. If now $h: A \rightarrow B$ is a homomorphism of R-algebras, then, of course, $h \circ \phi: M \rightarrow B$ is an R-module homomorphism of M into B. This observation leads us to pose the following universal problem.

Problem. *To choose A and ϕ so that given any R-module homomorphism $\psi: M \rightarrow B$ (B is an R-algebra) there shall exist a unique homomorphism $h: A \rightarrow B$, of R-algebras, such that $h \circ \phi = \psi$.*

It is clear that if (A, ϕ) and (A', ϕ') both solve this problem, then there will be inverse algebra-isomorphisms $\lambda: A \to A'$ and $\lambda': A' \to A$ such that $\lambda \circ \phi = \phi'$ and $\lambda' \circ \phi' = \phi$. Thus the problem has (essentially) at most one solution.

Theorem 1. *The universal problem possesses a solution. If (A, ϕ) is any solution, then*

(i) *$\phi: M \to A$ is an injection;*
(ii) *A is generated, as an R-algebra, by $\phi(M)$;*
(iii) *there is a grading $\{A_n\}_{n\in\mathbb{Z}}$ on A with $A_1 = \phi(M)$;*
(iv) *for $p \geq 1$, A_p is the p-th tensor power of M with $m_1 \otimes m_2 \otimes \cdots \otimes m_p = \phi(m_1)\phi(m_2)\ldots\phi(m_p)$;*
(v) *the structural homomorphism $R \to A$ maps R isomorphically onto A_0.*

Remarks. Before starting the proof it may be helpful if certain points are clarified. For instance it should be noted that, once (ii) has been established, Lemma 1 of Chapter 3 ensures that there is at most one grading on A that has the property described in (iii). To amplify (iv) we observe that the mapping of the p-fold product $M \times M \times \cdots \times M$ into A_p which takes $(m_1, m_2, \ldots, m_p)$ into $\phi(m_1)\phi(m_2)\ldots\phi(m_p)$ is multilinear. The theorem asserts that A_p and this mapping together solve the universal problem for multilinear mappings of $M \times M \times \cdots \times M$ (see Problem 1 of Chapter 1).

Proof. It suffices to exhibit one solution to the problem which satisfies (i)–(v). To this end, for $p \geq 1$, put $A_p = M \otimes M \otimes \cdots \otimes M$ where there are p factors. Thus $A_1 = M$. We also put $A_0 = R$ and define an R-module A by

$$A = \sum_{p\geq 0} A_p \quad \text{(d.s.)}.$$

Our immediate concern will be to turn A into an R-algebra.

For $p \geq 1$ and $q \geq 1$, Theorem 1 of Chapter 2 provides an isomorphism $A_p \otimes A_q \approx A_{p+q}$. Next, from Theorem 3 of the same chapter we obtain isomorphisms $A_0 \otimes A_q \approx A_q$ and $A_p \otimes A_0 \approx A_p$. (Note that when $p = q = 0$ the latter isomorphisms coincide.) Thus for every $p \geq 0$ and $q \geq 0$ we have an explicit isomorphism $A_p \otimes A_q \approx A_{p+q}$ and hence there is a bilinear mapping

$$\mu_{pq}: A_p \times A_q \to A_{p+q} \tag{4.1.1}$$

in which $\mu_{pq}(x_p, y_q)$ is the image of $x_p \otimes y_q$ under the isomorphism in question. Accordingly if $m_1, m_2, \ldots, m_p$ and $m_1', m_2', \ldots, m_q'$ belong to M, then

$$\begin{aligned}\mu_{pq}(m_1 \otimes \cdots \otimes m_p, m_1' \otimes \cdots \otimes m_q') \\ = m_1 \otimes \cdots \otimes m_p \otimes m_1' \otimes \cdots \otimes m_q'.\end{aligned} \tag{4.1.2}$$

On the other hand if $r \in A_0 = R$, then we have

$$\mu_{0q}(r, y_q) = ry_q \tag{4.1.3}$$

and

$$\mu_{p0}(x_p, r) = rx_p, \tag{4.1.4}$$

where now $x_p \in A_p$, $y_q \in A_q$ and $p \geq 0$, $q \geq 0$.

Suppose next that $p \geq 0$, $q \geq 0$, $t \geq 0$ and let x_p, y_q, z_t belong to A_p, A_q, A_t respectively. Using (4.1.2), (4.1.3) and (4.1.4) it is a straightforward matter to verify that

$$\mu_{p+q,t}(\mu_{pq}(x_p, y_q), z_t) = \mu_{p,q+t}(x_p, \mu_{qt}(y_p, z_t)). \tag{4.1.5}$$

We are now ready to define multiplication on A. Let $x, y \in A$. These elements have unique representations in the form $x = x_0 + x_1 + x_2 + \cdots$ and $y = y_0 + y_1 + y_2 + \cdots$, where x_n, y_n belong to A_n and the sums are essentially finite. The required product of x and y is then $\mu(x, y)$, where

$$\mu(x, y) = \sum_{p \geq 0, q \geq 0} \mu_{pq}(x_p, y_q).$$

It is readily checked that multiplication is distributive (on both sides) with respect to addition, and from (4.1.5) it follows that it is associative as well. Next 1_R belongs to A_0 and $\mu(1_R, y) = y$, $\mu(x, 1_R) = x$ by virtue of (4.1.3) and (4.1.4). Finally if $r \in R$, then

$$\mu(rx, y) = r\mu(x, y) = \mu(x, ry).$$

These observations taken together show that A is an (associative) R-algebra, and that 1_R, considered as an element of A_0, is its identity element.

For $n < 0$ put $A_n = 0$. Then $\{A_n\}_{n \in \mathbb{Z}}$ is a grading on A, and, since $M = A_1$, M is a submodule of A. Let $\phi: M \to A$ be the inclusion mapping. Evidently A and ϕ satisfy conditions (i)–(v). The proof will therefore be complete if we show that A and ϕ solve the problem which was enunciated at the beginning of the section.

Assume then that $\psi: M \to B$ is an R-linear mapping of M into an R-algebra B. For $p \geq 1$ there is a multilinear mapping of the p-fold product $M \times M \times \cdots \times M$ into B that takes $(m_1, m_2, \ldots, m_p)$ into $\psi(m_1)\psi(m_2) \ldots \psi(m_p)$, and this will induce a homomorphism $h_p: A_p \to B$, where

$$h_p(m_1 \otimes m_2 \otimes \cdots \otimes m_p) = \psi(m_1)\psi(m_2) \ldots \psi(m_p).$$

Since $A_0 = R$, we may take $h_0: A_0 \to B$ to be the structural homomorphism $R \to B$. But A is the direct sum of all the A_n with $n \geq 0$. Consequently there is an R-homomorphism $h: A \to B$ which agrees with h_n on A_n. Certainly h preserves identity elements. Also if x and y are homogeneous elements, then, using whichever of (4.1.2), (4.1.3) or (4.1.4) is appropriate, we can verify that $h(xy) = h(x)h(y)$. It follows that $h: A \to B$ is a homomorphism of R-algebras.

Finally for $x \in M = A_1$ we have $h(x) = \psi(x)$ and therefore $h \circ \phi = \psi$. But $M = A_1$ generates A as an R-algebra and so there can only be one algebra-homomorphism (of A into B) which when combined with ϕ gives ψ. Consequently the proof is complete.

These conclusions will now be cast into a different form. Suppose that (A, ϕ) solves our problem. Put

$$T(M) = A \tag{4.1.6}$$

and denote by $\{T_n(M)\}_{n \in \mathbb{Z}}$ the grading on $T(M)$ that results from Theorem 1. Since ϕ maps M isomorphically onto $T_1(M)$ we can use this isomorphism to identify the two modules. Thus

$$T_1(M) = M \tag{4.1.7}$$

so that $T(M)$ contains M as a submodule. Indeed by this device the notation is greatly simplified. We know that, as an R-algebra, $T(M)$ is generated by $T_1(M) = M$ so that the grading is non-negative. Also the structural homomorphism maps R isomorphically onto $T_0(M)$. Hence, *when it suits us*, we can identify $T_0(M)$ with R. Finally, if $\otimes$ is used to denote multiplication on $T(M)$, then, for $p \geq 1$, $T_p(M)$ is the p-th tensor power of M as defined in Section (1.3). It is because of this last property that $T(M)$ is called the *tensor algebra* of M.

Our next theorem merely restates, but now in the new terminology, that the tensor algebra solves the universal problem with which we started.

Theorem 2. *Let $T(M)$ be the tensor algebra of the R-module M, and let $\psi: M \to B$ be an R-linear mapping of M into an R-algebra B. Then ψ has a unique extension to an algebra-homomorphism of $T(M)$ into B.*

4.2 Functorial properties

Let $f: M \to N$ be a homomorphism of R-modules and let $T(M)$, $T(N)$ be the tensor algebras of M and N respectively. Since N is a submodule of $T(N)$, Theorem 2 shows that f has a unique extension to a homomorphism $T(f): T(M) \to T(N)$ of R-algebras. Note that if $m_1, m_2, \ldots, m_p$ belong to M, then

$$T(f)(m_1 \otimes m_2 \otimes \cdots \otimes m_p) = f(m_1) \otimes f(m_2) \otimes \cdots \otimes f(m_p). \tag{4.2.1}$$

Clearly if $N = M$ and f is the identity mapping of M, then $T(f)$ is the identity mapping of $T(M)$.

We next observe, as a consequence of (4.2.1), that a homogeneous element of degree $p \geq 1$ keeps its degree when $T(f)$ is applied. Moreover $T(f)$ maps $T_0(M)$ isomorphically onto $T_0(N)$. Consequently $T(f)$ is actually a homomorphism of *graded* algebras and therefore, for each $s \in \mathbb{Z}$, it induces a homomorphism

$$T_s(f): T_s(M) \to T_s(N) \tag{4.2.2}$$

of R-modules. In fact, as is shown by (4.2.1), when $p \geq 1$ we have

$$T_p(f) = f \otimes f \otimes \cdots \otimes f \quad (p \text{ factors}) \tag{4.2.3}$$

and to this we may add the observation that $T_0(f): T_0(M) \to T_0(N)$ is the identity mapping when we make the identifications $T_0(M) = R$ and $T_0(N) = R$.

Next suppose that we have R-homomorphisms $f: M \to N$ and $g: K \to M$. Evidently

$$T(f) \circ T(g) = T(f \circ g) \tag{4.2.4}$$

and hence, for each $s \in \mathbb{Z}$,

$$T_s(f) \circ T_s(g) = T_s(f \circ g). \tag{4.2.5}$$

Thus in the language of the Theory of Categories $T(M)$ is a covariant functor from R-modules to graded R-algebras. Note that if $f: M \to N$ is an isomorphism, then $T(f)$ is also an isomorphism and $T(f^{-1})$ is its inverse. This is because $T(f^{-1}) \circ T(f)$ and $T(f) \circ T(f^{-1})$ are identity mappings.

Now let K be a submodule of the R-module M. Since $T_1(M) = M$, K consists of homogeneous elements of $T(M)$ of degree one. Accordingly K generates a homogeneous two-sided ideal of $T(M)$.

Theorem 3. *Let $f: M \to N$ be a surjective homomorphism of R-modules and let K be its kernel. Then $T(f): T(M) \to T(N)$ is surjective and its kernel is the two-sided ideal that K generates in $T(M)$.*

Proof. It is clear from (4.2.1) that $T(f)$ is surjective. Let $I(K)$ be the two-sided ideal generated by K and put $I_p(K) = I(K) \cap T_p(M)$. Then, for $p \geq 1$, $I_p(K)$ is the submodule of $T_p(M)$ generated by all products $m_1 \otimes m_2 \otimes \cdots \otimes m_p$, where $m_1, m_2, \ldots, m_p$ belong to M and at least one m_i is in K. Since $T_p(f) = f \otimes f \otimes \cdots \otimes f$, it follows (Chapter 2, Theorem 4) that $I_p(K)$ is the kernel of $T_p(f)$. Consequently the kernel of $T(f)$ is

$$I_0(K) + I_1(K) + I_2(K) + \cdots = I(K)$$

as required.

We shall now give a second proof that the kernel of $T(f)$ is $I(K)$. This alternative proof has the advantage that it can be adapted to deal with similar situations that arise in the study of other algebras.

Let $h: T(M) \to T(M)/I(K)$ be the natural homomorphism of graded algebras. Since $T_1(M) = M$ and $I_1(K) = K$, the submodule of $T(M)/I(K)$ formed by the homogeneous elements of degree one can be identified with M/K, and then, in degree one, h induces the natural mapping $M \to M/K$. On the other hand, f induces an isomorphism of M/K onto N. Let $g: N \to M/K$ be the inverse of this isomorphism. We now have an R-module homomorphism $N \to T(M)/I(K)$ and this, by Theorem 2, extends to an algebra-homomorphism $T(N) \to T(M)/I(K)$. Let us combine the algebra-homomorphism $T(N) \to T(M)/I(K)$ with $T(f): T(M) \to T(N)$. The result is an algebra-homomorphism $T(M) \to T(M)/I(K)$ which maps m of $T_1(M)$ into $h(m)$. But $T_1(M)$ generates $T(M)$ as an R-algebra, so in combining the two homomorphisms we have recovered $h: T(M) \to T(M)/I(K)$. It follows that the kernel of $T(f)$ is contained in the kernel of h, that is to say it is contained in $I(K)$. The proof is now complete because the opposite inclusion is trivial.

4.3 The tensor algebra of a free module

In Section (4.3) we shall assume that R is non-trivial. (This is to avoid certain tiresome and unimportant complications.) Let M be a free R-module and B one of its bases. We wish to describe the structure of $T(M)$.

We know that $T_0(M)$ is a free R-module having the identity element of $T(M)$ as a base. Also, by Theorem 3 of Chapter 1, if $p \geq 1$, then $T_p(M)$ is a free R-module and it has as a base the set formed by all products $b_1 \otimes b_2 \otimes \cdots \otimes b_p$, where $b_i \in B$. Accordingly $T(M)$ is itself a free R-module and the totality of products $b_1 \otimes b_2 \otimes \cdots \otimes b_n$, where $b_i \in B$ and $n \geq 0$, constitutes a base. To complete the description of the algebra all we have to do is explain how two elements in our base are to be multiplied. But here there is no problem because the product of $b_1 \otimes b_2 \otimes \cdots \otimes b_n$ and $b'_1 \otimes b'_2 \otimes \cdots \otimes b'_\nu$ is simply the basis element $b_1 \otimes \cdots \otimes b_n \otimes b'_1 \otimes \cdots \otimes b'_\nu$. Thus $T(M)$ is what is called the *free algebra* generated by the set B.

The next theorem gives another characterization of $T(M)$ when M is free.

Theorem 4. *Let M be a free R-module with a base B and let $\phi: B \to A$ be a mapping of B into an R-algebra A. Then ϕ has a unique extension to a homomorphism $T(M) \to A$ of R-algebras.*

Remark. This result shows that $T(M)$ solves a certain universal problem involving mappings of B into an R-algebra. We leave the reader to make this precise.

Proof. The mapping $\phi: B \to A$ can be extended to an R-linear mapping $\bar{\phi}: M \to A$. Let $h: T(M) \to A$ be a homomorphism of R-algebras. Then h extends ϕ if and only if it extends $\bar{\phi}$. However, we know (Theorem 2) that there is precisely one h with the latter property.

Theorem 5. *Let P be a projective R-module. Then the tensor algebra $T(P)$ is projective as an R-module. Consequently (for all n) $T_n(P)$, since it is a direct summand of $T(P)$, is also projective.*

Proof. Choose an R-module Q so that $P \oplus Q = F$ (say) is a free R-module. Let $\sigma: P \to F$ be the inclusion mapping and $\pi: F \to P$ the projection onto the first summand. Then $T(F)$ is a free R-module (because F is free) and the mappings $T(\sigma): T(P) \to T(F)$ and $T(\pi): T(F) \to T(P)$ are R-linear. But $\pi \circ \sigma$ is the identity mapping of P and therefore $T(\pi) \circ T(\sigma)$ is the identity mapping of $T(P)$. The following lemma allows us to conclude that $T(P)$ is isomorphic to a direct summand of $T(F)$; and, since $T(F)$ is free, this means that $T(P)$ is projective.

Lemma 1. *Let $u: N \to M$ and $v: M \to N$ be homomorphisms of R-modules such that $v \circ u$ is the identity mapping of N. Then u is an injection, v is a surjection, and M is the direct sum of $u(N)$ and the kernel,* Ker v, *of v. In particular N is isomorphic to the direct summand $u(N)$ of M.*

Proof. It is obvious that u is an injection and v a surjection. Let $m \in M$. Since

$$m = (m - u(v(m))) + u(v(m))$$

and $m - u(v(m))$ is in Ker v, it follows that $M = \text{Ker } v + u(N)$. Now let $x \in u(N) \cap \text{Ker } v$. To complete the proof it suffices to show that $x = 0$. But $x = u(n)$ for some $n \in N$ and now

$$n = (v \circ u)(n) = v(x) = 0.$$

Accordingly $x = 0$ and the proof is complete.

4.4 Covariant extension of a tensor algebra

Let M be an R-module and

$$\omega: R \to S \tag{4.4.1}$$

a homomorphism of commutative rings. Our aim is to study the effect of the covariant extension associated with (4.4.1) on the tensor algebra of M. Since here we shall be concerned with more than one commutative ring, we shall embellish the symbol for the tensor algebra by writing $T_R(M)$ rather than just $T(M)$. This will help us to avoid certain ambiguities.

We know that $T_R(M)$ is a graded R-algebra and therefore (see Section (3.6)) its covariant extension $T_R(M) \otimes_R S$ is an S-algebra that is graded by the family $\{T_n(M) \otimes_R S\}_{n \in \mathbb{Z}}$ of S-submodules. The elements of degree one form the S-module $M \otimes_R S$ and this module generates $T_R(M) \otimes_R S$ as an S-algebra.

By Theorem 2 there is a homomorphism

$$\lambda: T_S(M \otimes_R S) \to T_R(M) \otimes_R S \tag{4.4.2}$$

of S-algebras which extends the inclusion mapping of $M \otimes_R S$ in $T_R(M) \otimes_R S$. (Here $T_S(M \otimes_R S)$ denotes the tensor algebra of the S-module $M \otimes_R S$.) Evidently λ preserves degrees and therefore it is a homomorphism of *graded* S-algebras. However, as the following theorem shows, much more is true.

Theorem 6. *The mapping*

$$\lambda: T_S(M \otimes_R S) \to T_R(M) \otimes_R S$$

(described above) is an isomorphism of graded S-algebras.

Proof. Any S-algebra, and in particular $T_S(M \otimes_R S)$, can be regarded as an R-algebra by leaving the ring structure unchanged and using (4.4.1) to turn it into an R-module. Now the mapping $M \to T_S(M \otimes_R S)$ which takes m into $m \otimes 1$ is R-linear. Accordingly, by Theorem 2, there is a homomorphism

$$\theta: T_R(M) \to T_S(M \otimes_R S)$$

of R-algebras such that $\theta(m) = m \otimes 1$ for all $m \in M$.

Consider the mapping $T_R(M) \times S \to T_S(M \otimes_R S)$ in which the image of (x, s) is $s\theta(x)$. This is a bilinear mapping of R-modules, and therefore it induces a homomorphism

$$\mu: T_R(M) \otimes_R S \to T_S(M \otimes_R S)$$

of R-modules which is such that $\mu(x \otimes s) = s\theta(x)$ for all $x \in T_R(M)$ and $s \in S$. It is now a simple matter to verify that μ is actually a homomorphism of S-algebras and that $\mu(m \otimes s) = m \otimes s$ for $m \in M$ and $s \in S$.

Consider $\mu \circ \lambda$ and $\lambda \circ \mu$. Each of these is a homomorphism of S-algebras and each leaves $M \otimes_R S$ elementwise fixed. It follows from this (and the fact that the S-algebras concerned are generated by $M \otimes_R S$) that $\mu \circ \lambda$ and $\lambda \circ \mu$ are identity mappings and hence that λ is an isomorphism. The proof is now complete.

In view of Theorem 6 we may write

$$T_S(M \otimes_R S) = T_R(M) \otimes_R S \tag{4.4.3}$$

and

$$T_n(M \otimes_R S) = T_n(M) \otimes_R S, \tag{4.4.4}$$

and we may note, in passing, that (4.4.4) is a special case of Theorem 9 of Chapter 2.

4.5 Derivations and skew derivations on a tensor algebra

Let M be an R-module. An R-linear mapping of M into R is called a *linear form* on M. If now D is a derivation on $T(M)$, then, since it is of degree -1, it induces an R-homomorphism $T_1(M) \to T_0(M)$. But $T_1(M) = M$ and we know that we can identify $T_0(M)$ with R. Thus D extends a linear form on M. Moreover it follows from Lemma 4 of Chapter 3 that two different derivations cannot extend the same linear form.

Similar considerations, this time based on Lemma 6 of Chapter 3, show that each skew derivation extends a linear form and that distinct skew derivations give rise to distinct linear forms. In the next two theorems it will be proved that $T(M)$ is fully endowed with both derivations and skew derivations.

Theorem 7. *Let f be a linear form on the R-module M. Then there is exactly one derivation on $T(M)$ which extends f.*

Proof. Suppose that $p \geq 1$. There is a multilinear mapping of the p-fold product $M \times M \times \cdots \times M$ into $T_{p-1}(M)$ in which $(m_1, m_2, \ldots, m_p)$ becomes

$$\sum_{i=1}^{p} f(m_i)(m_1 \otimes \cdots \otimes \hat{m}_i \otimes \cdots \otimes m_p). \tag{4.5.1}$$

(Here, as usual, the ˆ over m_i indicates that this factor is to be omitted.) Consequently there is induced an R-homomorphism D_p: $T_p(M) \to T_{p-1}(M)$, where $D_p(m_1 \otimes m_2 \otimes \cdots \otimes m_p)$ is the element (4.5.1). Note that $D_1 = f$.

Next the homomorphisms $D_1, D_2, D_3, \ldots$ give rise to an R-endomorphism D: $T(M) \to T(M)$, of degree -1, which agrees with D_p on $T_p(M)$. To complete the proof we show that D is a derivation.

Let $m_1, m_2, \ldots, m_p$ and $m_1', m_2', \ldots, m_q'$ belong to M and put $x = m_1 \otimes m_2 \otimes \cdots \otimes m_p$, $x' = m_1' \otimes m_2' \otimes \cdots \otimes m_q'$. Then

$$\begin{aligned} D(x \otimes x') &= D(m_1 \otimes \cdots \otimes m_p \otimes m_1' \otimes \cdots \otimes m_q') \\ &= \sum_{i=1}^{p} f(m_i)(m_1 \otimes \cdots \otimes \hat{m}_i \otimes \cdots \otimes m_p \otimes m_1' \otimes \cdots \otimes m_q') \\ &\quad + \sum_{j=1}^{q} f(m_j')(m_1 \otimes \cdots \otimes m_p \otimes m_1' \otimes \cdots \otimes \hat{m}_j' \otimes \cdots \otimes m_q') \\ &= (Dx) \otimes x' + x \otimes (Dx'). \end{aligned} \tag{4.5.2}$$

Since this relation extends by linearity to any two elements of $T(M)$, the proof is complete.

Theorem 8. *Let f be a linear form on the R-module M. Then there is one and only one skew derivation on $T(M)$ which extends f.*

Proof. This parallels the proof of Theorem 7. We first show that for each $p \geq 1$, there is an R-homomorphism Δ_p: $T_p(M) \to T_{p-1}(M)$, where

$$\Delta_p(m_1 \otimes m_2 \otimes \cdots \otimes m_p) = \sum_{i=1}^{p} (-1)^{i+1} f(m_i)(m_1 \otimes \cdots \otimes \hat{m}_i \otimes \cdots \otimes m_p).$$

These combine to yield an endomorphism Δ: $T(M) \to T(M)$, of degree -1, which extends f. Moreover, in place of (4.5.2) we obtain

$$\begin{aligned}\Delta(m_1 \otimes \cdots \otimes m_p \otimes m_1' \otimes \cdots \otimes m_q') \\ = \Delta(m_1 \otimes \cdots \otimes m_p) \otimes (m_1' \otimes \cdots \otimes m_q') \\ + (-1)^p (m_1 \otimes \cdots \otimes m_p) \otimes \Delta(m_1' \otimes \cdots \otimes m_q').\end{aligned}$$

This shows that Δ is a skew derivation and ends the proof.

4.6 Comments and exercises

We make here a few comments on the theory of tensor algebras and provide some exercises. As usual certain of these exercises have been marked with an asterisk; for these particular exercises solutions are provided in the next section.

It is clear that if an R-module M is generated by a given set of elements, then the same set of elements will also generate $T(M)$ as an R-algebra. It follows that if M is a *cyclic* R-module, then $T(M)$ can be generated by a single element. Consequently (see Exercise 1 of Chapter 3) the tensor algebra of a cyclic module is a homomorphic image of the polynomial ring $R[X]$ and, in particular, it is *commutative*. The first exercise builds on this observation.

Exercise 1*. *Let M be an R-module with the property that every finitely generated submodule is contained in a cyclic submodule. Show that its tensor algebra is commutative.*

For example, if R is an integral domain and F is its quotient field, then the tensor algebra of F (considered as an R-module) is commutative.

We remarked in Section (4.3) that if M is a free R-module with a base B, then $T(M)$ is the free R-algebra generated by B. (The notion of the free algebra generated by a set was enlarged upon in Section (3.8).) The next exercise is concerned with such an algebra in the special case where the ground ring R is an integral domain.

Exercise 2*. *Let R be an integral domain and M a free R-module. Further let x and y be non-zero elements of the tensor algebra $T(M)$. Show that the product of x and y in $T(M)$ is not zero.*

Once again let M be an R-module. In Section (3.8) we saw how a commutative, graded R-algebra can be formed from the matrices

$$\begin{bmatrix} r & m \\ 0 & r \end{bmatrix},$$

where $r \in R$ and $m \in M$. Let us denote this algebra by $\Gamma(M)$ and let $\Gamma_n(M)$ denote the submodule formed by the homogeneous elements of degree n. We recall that $\Gamma_0(M)$ consists of all the matrices

$$\begin{bmatrix} r & 0 \\ 0 & r \end{bmatrix}$$

whereas $\Gamma_1(M)$ is made up of the matrices

$$\begin{bmatrix} 0 & m \\ 0 & 0 \end{bmatrix}.$$

Finally $\Gamma_n(M)=0$ if $n \neq 0, 1$.

Consider the mapping $M \to \Gamma(M)$ in which m of M is mapping into

$$\begin{bmatrix} 0 & m \\ 0 & 0 \end{bmatrix}.$$

This mapping is R-linear and so it extends to a homomorphism

$$\lambda\colon T(M) \to \Gamma(M) \tag{4.6.1}$$

of R-algebras. Evidently λ preserves degrees and is surjective.

Exercise 3*. *Let $\lambda\colon T(M) \to \Gamma(M)$ be the homomorphism of graded R-algebras described in* (4.6.1). *Determine its kernel and show that λ is an isomorphism if and only if $M \otimes_R M = 0$.*

Exercise 4. *Let R be an integral domain and F its quotient field. Use Exercise* 3 *to give a simplified description of the tensor algebra of the R-module F/R.*

The next exercise is of general interest. Let $I \neq R$ be an ideal of R and let M be an R-module. Then $IT(M)$ is a homogeneous two-sided ideal of $T(M)$ and therefore $T(M)/IT(M)$ is a graded R-algebra. But I annihilates this algebra so that $T(M)/IT(M)=A$ (say) can be regarded as a graded R/I-algebra. Now the submodule A_1, of A, formed by the homogeneous elements of degree one is the image of $T_1(M)=M$, and the kernel of $T_1(M) \to A_1$ is IM. Hence the natural mapping $T(M) \to T(M)/IT(M)$ induces a homomorphism

$$M/IM \to T(M)/IT(M)$$

of R/I-modules and this in turn induces a homomorphism

$$\phi\colon T_{R/I}(M/IM) \to T(M)/IT(M) \tag{4.6.2}$$

of R/I-algebras. (Here by $T_{R/I}(M/IM)$ we mean the tensor algebra of M/IM considered as an R/I-module.)

Exercise 5*. *Show that the homomorphism*

$$\phi\colon T_{R/I}(M/IM) \to T(M)/IT(M)$$

of (4.6.2) *is an isomorphism of graded R/I-algebras.*

This result enables us to make the identification

$$T_{R/I}(M/IM) = T(M)/IT(M). \tag{4.6.3}$$

Our final comments concern generalized ordinary and skew derivations. These were defined in Section (3.8). First suppose that D is a generalized derivation of degree i on the tensor algebra of the R-module M. Then D induces an R-linear mapping of $T_1(M)=M$ into $T_{i+1}(M)$. Next $1_R \in T_0(M)$ and

$$D(1_R)=D(1_R 1_R)=D(1_R)+D(1_R)$$

so that $D(1_R)=0$. It follows that the mapping $T_0(M) \to T_i(M)$ induced by D is a null mapping. Again, if $p \geq 1$ and $m_1, m_2, \ldots, m_p$ belong to M, then it is easily seen (using induction on p) that

$$D(m_1 \otimes m_2 \otimes \cdots \otimes m_p) = \sum_{\nu=1}^{p} m_1 \otimes \cdots \otimes Dm_\nu \otimes \cdots \otimes m_p. \quad (4.6.4)$$

It follows that D is fully determined by its degree and its effect on $T_1(M)=M$.

Next assume that Δ is a generalized skew derivation on $T(M)$ of degree i. Everything that was said about D in the last paragraph applies to Δ except that (4.6.4) has to be replaced by

$$\Delta(m_1 \otimes m_2 \otimes \cdots \otimes m_p)$$
$$= \sum_{\nu=1}^{p} (-1)^{(\nu+1)i} m_1 \otimes \cdots \otimes \Delta m_\nu \otimes \cdots \otimes m_p. \quad (4.6.5)$$

The next exercise generalizes Theorems 7 and 8.

Exercise 6*. *Let M be an R-module, let i be a given integer, and let $f: M \to T_{i+1}(M)$ be an R-linear mapping. Show that there is exactly one generalized derivation of degree i and exactly one generalized skew derivation of degree i, on $T(M)$, that extend f.*

4.7 Solutions to selected exercises

Exercise 1. *Let M be an R-module with the property that every finitely generated submodule is contained in a cyclic submodule. Show that its tensor algebra is commutative.*

Solution. Let ξ, ξ' belong to $T(M)$. We wish to prove that $\xi\xi'=\xi'\xi$ and this will follow with full generality if we can establish it on the assumption that $\xi=m_1 \otimes m_2 \otimes \cdots \otimes m_p$, $\xi'=m_1' \otimes m_2' \otimes \cdots \otimes m_q'$, where the m_i and m_j' are in M. Now, by hypothesis, there exists a cyclic submodule N, of M, containing all the m_i and the m_j'. Let ϕ: $T(N) \to T(M)$ be the algebra-homomorphism that extends the inclusion mapping of N into M. Then there exist x and x' in $T(N)$ such that $\phi(x)=\xi$ and $\phi(x')=\xi'$ and therefore $\phi(xx')=\xi\xi'$ and $\phi(x'x)=\xi'\xi$. But $T(N)$ is commutative because N is cyclic. Hence $\xi\xi'=\xi'\xi$ as required.

Exercise 2. *Let R be an integral domain and M a free R-module. Further let x and y be non-zero elements of the tensor algebra $T(M)$. Show that the product of x and y in $T(M)$ is not zero.*

Solution. We can write

$$x = x_0 + x_1 + x_2 + \cdots + x_h \quad (x_h \neq 0),$$

$$y = y_0 + y_1 + y_2 + \cdots + y_k \quad (y_k \neq 0),$$

where $x_i \in T_i(M)$ and $y_j \in T_j(M)$, and it will suffice to prove that the product of x_h and y_k is not zero. Equally well we may assume for the rest of the proof that x belongs to $T_h(M)$ and y to $T_k(M)$. Before proceeding note that under the canonical isomorphism

$$T_h(M) \otimes T_k(M) \approx T_{h+k}(M)$$

the element of $T_{h+k}(M)$ that corresponds to $x \otimes y$ is precisely the product of x and y in $T(M)$.

Put $U = T_h(M)$ and $V = T_k(M)$. These are free R-modules because M is free. Let $u_1, u_2, \ldots, u_p$ be a base for U and $v_1, v_2, \ldots, v_q$ a base for V. Then (see Chapter 1, Theorem 3) the elements $u_i \otimes v_j$ form a base for $U \otimes V$.

Next, we can write $x = a_1 u_1 + a_2 u_2 + \cdots + a_p u_p$ and $y = b_1 v_1 + b_2 v_2 + \cdots + b_q v_q$, where a_i, b_j belong to R, and then

$$x \otimes y = \sum_{i,j} a_i b_j (u_i \otimes v_j).$$

But $x \neq 0$ and $y \neq 0$. Hence we can find i and j so that $a_i \neq 0$ and $b_j \neq 0$ and then we have $a_i b_j \neq 0$ because R is an integral domain. However, the products $u_r \otimes v_s$ are linearly independent over R and therefore it follows that $x \otimes y \neq 0$. This completes the solution.

Exercise 3. *Let $\lambda: T(M) \to \Gamma(M)$ be the homomorphism of graded R-algebras described in* (4.6.1). *Determine its kernel and show that λ is an isomorphism if and only if $M \otimes_R M = 0$.*

Solution. Let $m_1, m_2, \ldots, m_p$ belong to M. Then $\lambda(m_1 \otimes m_2 \otimes \cdots \otimes m_p)$ is the product of the matrices

$$\begin{bmatrix} 0 & m_i \\ 0 & 0 \end{bmatrix} \quad (i = 1, 2, \ldots, p)$$

and this is zero if $p \geq 2$. Thus λ maps all of $T_2(M)$, $T_3(M)$, $T_4(M), \ldots$ into zero.

Let x belong to $T(M)$ and let

$$x = x_0 + x_1 + x_2 + \cdots$$

be its decomposition into homogeneous components. Then $x_0 \in R$, $x_1 \in M$ and

$$\lambda(x)=\lambda(x_0)+\lambda(x_1)=\begin{bmatrix} x_0 & x_1 \\ 0 & x_0 \end{bmatrix}.$$

It follows that

$$\text{Ker }\lambda=\sum_{n\geq 2} T_n(M).$$

Of course, since λ is surjective it is an isomorphism if and only if Ker $\lambda=0$.

Now if λ is an isomorphism, then $M \otimes M = T_2(M)=0$. On the other hand, if $M \otimes M=0$, then $M \otimes M \otimes \cdots \otimes M=0$ provided there are at least two factors (see Chapter 2, Theorem 1). Consequently if $M \otimes M=0$, then $T_n(M)=0$ for all $n\geq 2$ and therefore Ker $\lambda=0$. This completes the solution.

Exercise 5. *Show that the homomorphism*

$$\phi\colon T_{R/I}(M/IM)\to T(M)/IT(M)$$

of (4.6.2) *is an isomorphism of graded R/I-algebras.*

Solution. It is clear that ϕ preserves the degrees of homogeneous elements so that it is enough to construct an algebra-homomorphism, in the reverse direction, which when combined with ϕ (in either order) produces an identity mapping.

Let $h\colon T(M)\to T(M)/IT(M)$ be the natural homomorphism and for m in M let $\bar{m}$ denote its natural image in M/IM. Next, for the moment, let us regard the R/I-algebra $T_{R/I}(M/IM)$ as an R-algebra. Then the R-linear mapping $M\to T_{R/I}(M/IM)$ which takes m into $\bar{m}$ extends to a homomorphism $T(M)\to T_{R/I}(M/IM)$ of R-algebras. But this vanishes on $IT(M)$ and so it induces a homomorphism

$$\psi\colon T(M)/IT(M)\to T_{R/I}(M/IM).$$

This is a homomorphism of R-algebras and of R/I-algebras and it satisfies $\psi(h(m))=\bar{m}$ for all m in M. Now $\phi(\bar{m})=h(m)$ by construction so that $\psi(\phi(\bar{m}))=\bar{m}$ and $\phi(\psi(h(m)))=h(m)$. But the elements $\bar{m}$ generate the R/I-algebra $T_{R/I}(M/IM)$ and, because M generates $T(M)$, the elements $h(m)$ generate the R/I-algebra $T(M)/IT(M)$. Thus $\psi\circ\phi$ and $\phi\circ\psi$ must be identity mappings and with this the solution is complete.

Exercise 6. *Let M be an R-module, let i be a given integer, and let $f\colon M\to T_{i+1}(M)$ be an R-linear mapping. Show that there is exactly one generalized derivation of degree i and exactly one generalized skew derivation of degree i, on $T(M)$, that extend f.*

Solution. We shall only establish the assertion concerning generalized skew derivations because the other assertion can be dealt with similarly, the details being somewhat simpler in the case of ordinary derivations.

Suppose that $p \geq 1$. The mapping of the p-fold product $M \times M \times \cdots \times M$ into $T_{i+p}(M)$ which takes $(m_1, m_2, \ldots, m_p)$ into

$$\sum_{\mu=1}^{p} (-1)^{(\mu+1)i} m_1 \otimes \cdots \otimes f(m_\mu) \otimes \cdots \otimes m_p$$

is multilinear and therefore it induces an R-linear mapping

$$\Delta_p: T_p(M) \rightarrow T_{p+i}(M)$$

which is such that

$$\Delta_p(m_1 \otimes m_2 \otimes \cdots \otimes m_p) = \sum_{\mu=1}^{p} (-1)^{(\mu+1)i} m_1 \otimes \cdots \otimes f(m_\mu) \otimes \cdots \otimes m_p.$$

Next we define $\Delta_0: T_0(M) \rightarrow T_i(M)$ to be the null homomorphism. Then $\Delta_0, \Delta_1, \Delta_2, \ldots$ can be combined to give an R-linear mapping $\Delta: T(M) \rightarrow T(M)$ of degree i which agrees with Δ_n on $T_n(M)$ for all $n \geq 0$. Since $\Delta_1 = f$, Δ extends f. Now to show that Δ is a generalized skew derivation we have to check that, for x and y in $T(M)$, we have

$$\Delta(x \otimes y) = (\Delta x) \otimes y + J^i(x) \otimes (\Delta y),$$

where J is the main involution (see Section (3.8)). However, it suffices to do the checking when x and y are homogeneous elements whose degrees are strictly positive.

Suppose then that $x \in T_p(M)$ and $y \in T_q(M)$, where $p \geq 1$ and $q \geq 1$. We have to show that

$$\Delta(x \otimes y) = (\Delta x) \otimes y + (-1)^{ip} x \otimes (\Delta y).$$

But now we can specialize still further and suppose that $x = m_1 \otimes m_2 \otimes \cdots \otimes m_p$ and $y = m'_1 \otimes m'_2 \otimes \cdots \otimes m'_q$, where m_i and m'_j are elements of M. On this understanding we have

$$\begin{aligned} \Delta(x \otimes y) = &\sum_{\mu=1}^{p} (-1)^{(\mu+1)i} m_1 \otimes \cdots \otimes f(m_\mu) \otimes \cdots \otimes m_p \otimes m'_1 \otimes \cdots \otimes m'_q \\ &+ \sum_{\nu=1}^{q} (-1)^{(p+\nu+1)i} m_1 \otimes \cdots \otimes m_p \otimes m'_1 \otimes \cdots \otimes f(m'_\nu) \otimes \cdots \otimes m'_q \\ &= (\Delta x) \otimes y + (-1)^{ip} x \otimes (\Delta y) \end{aligned}$$

which is what we require. This establishes that there is a generalized skew derivation of degree i that extends f. That there can be at most one such skew derivation was already noted in Section (4.6).

5

The exterior algebra of a module

General remarks

The exterior algebra is one of the most interesting and useful of the algebras that can be derived from a module. As we shall see it is an anticommutative algebra having intimate connections with the theory of determinants. Our aim in this chapter will be to establish all its main properties; and in so doing we shall follow a pattern which can be used again, with only small modifications, to study a related algebra, namely the symmetric algebra of a module.

As usual R and S are reserved to denote commutative rings with an identity element, and we again allow ourselves the freedom to omit the suffix from the tensor symbol when there is no uncertainty concerning the ground ring. Algebras are understood to be associative and to have an identity element; and homomorphisms of rings and algebras are required to preserve identity elements. Finally $T(M)$ denotes the tensor algebra of an R-module M.

5.1 The exterior algebra

Let M be an R-module. We propose to define its exterior algebra, and, as in the case of the tensor algebra, we shall do this by means of a universal problem. However, because we are now in a position to make use of the properties of tensor algebras, the details on this occasion will be much simpler.

Suppose that $\phi: M \to A$ is an R-linear mapping of M into an R-algebra A, and suppose that $(\phi(m))^2 = 0$ for all $m \in M$. If now $h: A \to B$ is a homomorphism of R-algebras, then $h \circ \phi$ is an R-linear mapping of M into B with the property that, for every $m \in M$, the square of $(h \circ \phi)(m)$ is zero.

Problem. *To choose A and ϕ so that given any R-linear mapping $\psi: M \to B$*

(B is an R-algebra) such that $(\psi(m))^2=0$ for all $m \in M$, there shall exist a unique homomorphism $h: A \to B$, of R-algebras, such that $h \circ \phi = \psi$.

Evidently the problem has at most one solution. More precisely, if (A, ϕ) and (A', ϕ') both solve the problem, then there exist inverse isomorphisms $\lambda: A \to A'$ and $\lambda': A' \to A$, of R-algebras, such that $\lambda \circ \phi = \phi'$ and $\lambda' \circ \phi' = \phi$. The next theorem shows *inter alia* that there is always a solution.

Theorem 1. *The above universal problem possesses a solution. If (A, ϕ) is any solution, then*

(i) *$\phi: M \to A$ is an injection;*
(ii) *A is generated, as an R-algebra, by $\phi(M)$;*
(iii) *there is a grading $\{A_n\}_{n \in \mathbb{Z}}$ on A such that $A_1 = \phi(M)$;*
(iv) *for each $p \geq 1$, A_p is the p-th exterior power of M where, for $m_1, m_2, \ldots, m_p$ in M, we have $m_1 \wedge m_2 \wedge \cdots \wedge m_p = \phi(m_1)\phi(m_2) \ldots \phi(m_p)$;*
(v) *the structural homomorphism $R \to A$ maps R isomorphically onto A_0.*

Remark. As in the case of the tensor algebra, condition (ii) ensures that there can only be one grading with the property described in (iii).

Proof. It is sufficient to construct one solution of the universal problem that has all the five properties. To this end let $T(M)$ be the tensor algebra of M and let m belong to $M = T_1(M)$. Then $m \otimes m$ belongs to $T_2(M)$ and therefore such elements generate a homogeneous two-sided ideal, $J(M)$ say, in $T(M)$. Let $J_s(M) = J(M) \cap T_s(M)$. Then $J_0(M) = 0$ and $J_1(M) = 0$. Furthermore $J_n(M) = 0$ for all $n < 0$.

Since $J(M)$ is homogeneous, the factor algebra $T(M)/J(M) = A$ (say) has a grading $\{A_s\}_{s \in \mathbb{Z}}$ in which A_s is the image of $T_s(M)$. Bearing in mind that $T_1(M) = M$, we let $\phi: M \to A$ be the homomorphism induced by the natural mapping $T(M) \to T(M)/J(M)$. Then A and ϕ satisfy (i) because $J_1(M) = 0$. Since $T_1(M)$ generates $T(M)$ as an algebra, $\phi(M)$ generates A as an algebra so that (ii) holds as well. Evidently (iii) is satisfied. Again $T_0(M)$ is mapped isomorphically onto A_0 and therefore the requirement (v) is also met.

Now suppose that $p \geq 1$. The natural mapping of $T(M)$ onto A induces a surjective homomorphism of $T_p(M)$ onto A_p and the kernel of this homomorphism is $J_p(M)$. But, by the definition of $J(M)$, $J_p(M)$ is the submodule of $T_p(M)$ generated by all products $m_1 \otimes m_2 \otimes \cdots \otimes m_p$ with $m_i = m_{i+1}$ for some i. That condition (iv) holds therefore follows from Theorem 5 of Chapter 1.

It has thus been shown that A and ϕ satisfy the five conditions. To complete the proof we must now verify that they solve the universal problem. Let $m \in M$. Then $m \otimes m \in J(M)$ and therefore $(\phi(m))^2 = 0$.

Finally suppose that $\psi: M \to B$ is an R-linear mapping (of M into an R-algebra B) which is such that $(\psi(m))^2 = 0$. Then (Chapter 4, Theorem 2), ψ extends to an algebra-homomorphism $T(M) \to B$ which, it is clear, vanishes on $J(M)$. Accordingly there is induced a homomorphism $h: A \to B$ of R-algebras which satisfies $(h \circ \phi)(m) = \psi(m)$ for all $m \in M$. This shows that $h \circ \phi = \psi$ and now (ii) ensures that there is only one algebra-homomorphism with this property. Thus the proof is complete.

We next introduce some convenient terminology. Let (A, ϕ) solve our universal problem. Put

$$E(M) = A \tag{5.1.1}$$

and let $\{E_n(M)\}_{n \in \mathbb{Z}}$ be the grading referred to in Theorem 1. Since ϕ maps M isomorphically onto $E_1(M)$ we can use this fact to make the identification

$$E_1(M) = M. \tag{5.1.2}$$

In this way M becomes a submodule of $E(M)$ and, as we know, it generates $E(M)$ as an R-algebra. The grading on $E(M)$ is, of course, non-negative and the structural homomorphism $R \to E(M)$ maps R isomorphically onto $E_0(M)$. It follows that we may make the further identification

$$E_0(M) = R \tag{5.1.3}$$

when it is convenient to do so.

The symbol $\wedge$ will be used for multiplication on $E(M)$. This secures that, for $p \geq 1$, $E_p(M)$ is the p-th exterior power of M according to the definition given in Section (1.4). Note that

$$m \wedge m = 0 \tag{5.1.4}$$

for all m in $M = E_1(M)$.

The algebra $E(M)$ is called the *exterior algebra* of the module M. The following theorem simply records the fact that it solves the universal problem with which we started.

Theorem 2. *Let $E(M)$ be the exterior algebra of the R-module M and let $\psi: M \to B$ be an R-linear mapping, of M into an R-algebra B, such that $(\psi(m))^2 = 0$ for all $m \in M$. Then ψ has a unique extension to a homomorphism $E(M) \to B$ of R-algebras.*

Before we proceed to investigate the properties of the exterior algebra, let us make quite explicit its connection with the tensor algebra. By Theorem 2 of Chapter 4, the inclusion mapping $M \to E(M)$ has a unique extension to a homomorphism $T(M) \to E(M)$ of R-algebras. This algebra-homomorphism is *surjective*. We call it the *canonical homomorphism* of $T(M)$ onto $E(M)$. Now both algebras are generated by M and the canonical homomorphism leaves the elements of M fixed. It follows that the degree of a homogeneous element is preserved and therefore $T(M) \to E(M)$ is a homomorphism of *graded* algebras. In particular for each $p \in \mathbb{Z}$ there is induced a homomorphism

$$T_p(M) \to E_p(M) \tag{5.1.5}$$

of R-modules. Of course for $p=0$ this is an isomorphism and for $p=1$ it is the identity mapping. Suppose next that $p \geq 1$ and let $m_1, m_2, \ldots, m_p$ belong to M. Then in (5.1.5) $m_1 \otimes m_2 \otimes \cdots \otimes m_p$ is mapped into $m_1 \wedge m_2 \wedge \cdots \wedge m_p$ and therefore (5.1.5) is the canonical homomorphism of the p-th tensor power onto the p-th exterior power according to the definition given in Section (1.4).

Theorem 3. *The exterior algebra $E(M)$ of the R-module M is anticommutative.*

Proof. Since $m \wedge m = 0$ for all $m \in M$, the theorem follows from Lemma 3 of Chapter 3 and the properties of $E(M)$ that have already been noted.

5.2 Functorial properties

Let $f: M \to N$ be a homomorphism of R-modules. We can regard N as a submodule of the exterior algebra $E(N)$ and then $f(m) \wedge f(m) = 0$ for all $m \in M$. It follows (Theorem 2) that f has a unique extension to a homomorphism

$$E(f): E(M) \to E(N) \tag{5.2.1}$$

of R-algebras. This is such that if $m_1, m_2, \ldots, m_p$ belong to M, then

$$E(f)(m_1 \wedge m_2 \wedge \cdots \wedge m_p) = f(m_1) \wedge f(m_2) \wedge \cdots \wedge f(m_p). \tag{5.2.2}$$

Moreover, since M consists of the elements of degree one in $E(M)$, the homomorphism $E(f)$ preserves degrees. Thus $E(f): E(M) \to E(N)$ is a homomorphism of *graded* algebras and therefore, for each $p \in \mathbb{Z}$, it induces a homomorphism

$$E_p(f): E_p(M) \to E_p(N) \tag{5.2.3}$$

of R-modules. Of course $E_0(f)$ is an isomorphism and, when $p \geq 1$,

$$E_p(f)(m_1 \wedge m_2 \wedge \cdots \wedge m_p) = f(m_1) \wedge f(m_2) \wedge \cdots \wedge f(m_p). \tag{5.2.4}$$

Naturally when $N = M$ and f is its identity mapping, $E(f)$ is the identity mapping of $E(M)$.

Next suppose that in addition to $f: M \to N$ we are given a second homomorphism $g: K \to M$ of R-modules. Then

$$E(f) \circ E(g) = E(f \circ g) \tag{5.2.5}$$

and also

$$E_p(f) \circ E_p(g) = E_p(f \circ g) \tag{5.2.6}$$

for all $p \in \mathbb{Z}$. In particular the exterior algebra provides a second example of a covariant functor from R-modules to graded R-algebras. It follows, as in all such situations, that if f is an isomorphism of modules, then $E(f)$ is an isomorphism of algebras and $E(f^{-1})$ is its inverse.

At this point it is convenient to mention a notation which is very commonly used in the theory of exterior algebras and of which the reader should be aware. This is the *wedge notation* which we have already employed to describe multiplication. When used more extensively we put

$$\bigwedge M = E(M), \tag{5.2.7}$$

$$\bigwedge^p M = E_p(M), \tag{5.2.8}$$

$$\bigwedge f = E(f), \tag{5.2.9}$$

and

$$\bigwedge^p f = E_p(f). \tag{5.2.10}$$

Accordingly (5.2.5) and (5.2.6) become $(\bigwedge f) \circ (\bigwedge g) = \bigwedge (f \circ g)$ and $(\bigwedge^p f) \circ (\bigwedge^p g) = \bigwedge^p (f \circ g)$ respectively. However, since we aim to develop the theories of several algebras along parallel lines, we shall keep to a uniform notation. This will help a little when making comparisons.

Theorem 4. *Let $f: M \to N$ be a surjective homomorphism of R-modules and let K be its kernel. Then $E(f)$: $E(M) \to E(N)$ is a surjective homomorphism of graded algebras whose kernel is the two-sided ideal which K generates in $E(M)$.*

Remark. Since M is a submodule of $E(M)$ so too is K. It is therefore meaningful to speak of the two-sided ideal which K generates in $E(M)$.

Proof. The corresponding result for tensor algebras is Theorem 3 of Chapter 4. Two proofs were given of that result and, of these, the second one can be readily adapted. The details are left to the reader.

We now add a few remarks that are relevant to the study of the exterior algebra of a *finitely generated* module.

Let $e_1, e_2, \ldots, e_n$ belong to the R-module M and let elements $u_1, u_2, \ldots, u_r$ of M satisfy relations

$$u_s = e_1 c_{1s} + e_2 c_{2s} + \cdots + e_n c_{ns} \quad (s = 1, 2, \ldots, r). \tag{5.2.11}$$

(Here the coefficients c_{ij} are in R and, for convenience, we have written them on the right-hand side of the module elements which they multiply.) For integers $j_1, j_2, \ldots, j_r$, all between 1 and n, we put

$$C_{j_1 j_2 \ldots j_r} = \begin{vmatrix} c_{j_1 1} & c_{j_1 2} & \cdots & c_{j_1 r} \\ c_{j_2 1} & c_{j_2 2} & \cdots & c_{j_2 r} \\ \vdots & \vdots & \vdots & \vdots \\ c_{j_r 1} & c_{j_r 2} & \cdots & c_{j_r r} \end{vmatrix}. \tag{5.2.12}$$

Lemma 1. *Let the situation be as described above. Then*

$$u_1 \wedge u_2 \wedge \cdots \wedge u_r = \sum (e_{j_1} \wedge e_{j_2} \wedge \cdots \wedge e_{j_r}) C_{j_1 j_2 \ldots j_r},$$

where the sum is taken over all sequences $(j_1, j_2, \ldots, j_r)$ *which satisfy* $1 \leq j_1 < j_2 < \cdots < j_r \leq n$. *In particular* $u_1 \wedge u_2 \wedge \cdots \wedge u_r = 0$ *if* $r > n$.

Proof. We have

$$\begin{aligned} u_1 \wedge u_2 \wedge \cdots \wedge u_r &= (e_1 c_{11} + \cdots + e_n c_{n1}) \wedge (e_1 c_{12} + \cdots + e_n c_{n2}) \\ &\quad \wedge \cdots \wedge (e_1 c_{1r} + \cdots + e_n c_{nr}) \\ &= \sum (e_{i_1} \wedge e_{i_2} \wedge \cdots \wedge e_{i_r}) c_{i_1 1} c_{i_2 2} \ldots c_{i_r r}, \end{aligned} \tag{5.2.13}$$

where the summation is taken over *all* sequences $(i_1, i_2, \ldots, i_r)$ of r integers lying between 1 and n. However, $e_{i_1} \wedge e_{i_2} \wedge \cdots \wedge e_{i_r} = 0$ if the sequence $(i_1, i_2, \ldots, i_r)$ contains a repetition. Hence from now on we may assume that $r \leq n$.

Suppose that $1 \leq j_1 < j_2 < \cdots < j_r \leq n$ and let $(i_1, i_2, \ldots, i_r)$ be a permutation of $(j_1, j_2, \ldots, j_r)$. Then, because $e_p \wedge e_q = -e_q \wedge e_p$, we have

$$e_{i_1} \wedge e_{i_2} \wedge \cdots \wedge e_{i_r} = \pm e_{j_1} \wedge e_{j_2} \wedge \cdots \wedge e_{j_r},$$

where the plus (respectively minus) sign is to be taken if $(i_1, i_2, \ldots, i_r)$ is an even (respectively odd) permutation of $(j_1, j_2, \ldots, j_r)$. Accordingly the permutations of $(j_1, j_2, \ldots, j_r)$ contribute to (5.2.13) an amount equal to $(e_{j_1} \wedge e_{j_2} \wedge \cdots \wedge e_{j_r}) C_{j_1 j_2 \ldots j_r}$. The lemma follows.

As an application of the lemma we have

Theorem 5. *Let the R-module M be generated by* $e_1, e_2, \ldots, e_n$. *Then* $E_r(M) = 0$ *if* $r > n$. *If* $1 \leq r \leq n$, *then* $E_r(M)$ *is generated, as an R-module, by the elements* $e_{j_1} \wedge e_{j_2} \wedge \cdots \wedge e_{j_r}$, *where* $(j_1, j_2, \ldots, j_r)$ *is a typical sequence of r integers satisfying* $1 \leq j_1 < j_2 < \cdots < j_r \leq n$.

5.3 The exterior algebra of a free module

Throughout Section (5.3) we shall assume that the ring R is *non-trivial.*

Let M be a free R-module. If $p \geq 1$, then $E_p(M)$ is free by Theorem 6 of Chapter 1. Moreover $E_0(M)$ is free because it is isomorphic to R. But

$$E(M) = \sum_{n \geq 0} E_n(M) \text{ (d.s.)}$$

so that $E(M)$ too is a free R-module.

Theorem 6. *Let P be a projective R-module. Then* $E(P)$, *considered as an R-module, is projective. Consequently, for all s,* $E_s(P)$ *(since it is a direct summand of* $E(P)$*) is a projective module.*

Proof. The corresponding result for tensor algebras is Theorem 5 of Chapter 4. The argument used there, with only trivial modifications, also yields the desired result on this occasion. The details are left to the reader.

Let M be a free R-module and let B be a base for M. If B is *infinite*, then, by Theorem 6 of Chapter 1, all the exterior powers $E_0(M), E_1(M), E_2(M), \ldots$ are non-zero free R-modules.

Now consider what happens when B consists of a finite number of elements, say $u_1, u_2, \ldots, u_n$, where $n \geq 0$. By Theorem 5, $E_r(M) = 0$ if $r > n$. On the other hand, if $1 \leq r \leq n$, then Theorem 6 of Chapter 1 shows that $E_r(M)$ has a base consisting of the products $u_{j_1} \wedge u_{j_2} \wedge \cdots \wedge u_{j_r}$, where $(j_1, j_2, \ldots, j_r)$ is a sequence of integers satisfying $1 \leq j_1 < j_2 < \cdots < j_r \leq n$. Thus $E_r(M)$ has a base consisting of $\binom{n}{r}$ elements and, in particular, $E_r(M) \neq 0$. Of course $E_0(M) \neq 0$ because it is isomorphic to R.

Theorem 7. *Let M be a free R-module and B be a base of M. Then the number of elements in B is the upper bound of all integers k such that $E_k(M) \neq 0$.*

This is an immediate consequence of our previous remarks.

Corollary. *Let B and B' be bases of the free module M. Then either B and B' are both infinite, or they are both finite and contain the same number of elements.*

The number of elements in a base of a free R-module M is called the *rank* of M and it will be denoted by $\operatorname{rank}_R(M)$. Thus $\operatorname{rank}_R(M)$ is either a non-negative integer or else it is 'plus infinity'. If $\operatorname{rank}_R(M) = \infty$, then $\operatorname{rank}_R(E_0(M)) = 1$ and $\operatorname{rank}_R(E_p(M)) = \infty$ for all $p \geq 1$. On the other hand if $\operatorname{rank}_R(M) = n$, then

$$\operatorname{rank}_R(E_s(M)) = \binom{n}{s} \tag{5.3.1}$$

for all $s \geq 0$.

Let F be a free R-module of rank n and let $f: F \to F$ be an R-endomorphism of F. Then $E_n(f)$ is an endomorphism of $E_n(F)$ and, by (5.3.1), $E_n(F)$ has a base consisting of a single element. Consequently there exists $\alpha \in R$ such that $E_n(F)$: $E_n(F) \to E_n(F)$ is the homothety α and, moreover, α is uniquely determined. This scalar α is called the *determinant* of f and it is denoted by $\operatorname{Det}(f)$. Hence

$$f(x_1) \wedge f(x_2) \wedge \cdots \wedge f(x_n) = \operatorname{Det}(f)(x_1 \wedge x_2 \wedge \cdots \wedge x_n), \tag{5.3.2}$$

where $x_1, x_2, \ldots, x_n$ are any n elements of F.

Suppose now that $e_1, e_2, \ldots, e_n$ is a base of F. Then

$$f(e_r) = \sum_{i=1}^{r} e_i c_{ir} \quad (r = 1, 2, \ldots, n), \tag{5.3.3}$$

where

$$C=\begin{bmatrix} c_{11} & c_{12} & \cdots & c_{1n} \\ c_{21} & c_{22} & \cdots & c_{2n} \\ \vdots & \vdots & \vdots\vdots & \vdots \\ c_{n1} & c_{n2} & \cdots & c_{nn} \end{bmatrix} \tag{5.3.4}$$

is the matrix of f with respect to the given base. By Lemma 1,

$$f(e_1)\wedge f(e_2)\wedge\cdots\wedge f(e_n)=\operatorname{Det}(C)(e_1\wedge e_2\wedge\cdots\wedge e_n). \tag{5.3.5}$$

Since $e_1\wedge e_2\wedge\cdots\wedge e_n$ is a base for $E_n(F)$ we see, from (5.3.2), that $\operatorname{Det}(f)=\operatorname{Det}(C)$. Thus the determinant of f equals the determinant of any of its representative matrices. Again if $g: F\to F$ is also an R-endomorphism of F, then from $E_n(g\circ f)=E_n(g)\circ E_n(f)$ it follows that $\operatorname{Det}(g\circ f)=\operatorname{Det}(g)\operatorname{Det}(f)$. When this is expressed in terms of representative matrices for f and g, we obtain, of course, the usual formula for the determinant of the product of two square matrices.

Once again let $e_1, e_2, \ldots, e_n$ be a base of F, let f be an R-endomorphism of F, and let $f(e_r)$ be given by (5.3.3). Suppose that $1<p<n$. In what follows we shall use $K=(k_1, k_2, \ldots, k_p)$ (respectively $J=(j_1, j_2, \ldots, j_p)$) to denote a sequence of p integers satisfying $1\le k_1<k_2<\cdots<k_p\le n$ (respectively $1\le j_1<j_2<\cdots<j_p\le n$), and K' (respectively J') will stand for the sequence obtained from $(1, 2, 3, \ldots, n)$ by deleting the terms that belong to K (respectively J).

Put $x_i=f(e_i)$, $x_K=x_{k_1}\wedge x_{k_2}\wedge\cdots\wedge x_{k_p}$, and $e_J=e_{j_1}\wedge e_{j_2}\wedge\cdots\wedge e_{j_p}$. Then, by Lemma 1,

$$x_K=\sum_J e_J C_{JK},$$

where

$$C_{JK}=\begin{vmatrix} c_{j_1k_1} & c_{j_1k_2} & \cdots & c_{j_1k_p} \\ c_{j_2k_1} & c_{j_2k_2} & \cdots & c_{j_2k_p} \\ \vdots & \vdots & \vdots\vdots & \vdots \\ c_{j_pk_1} & c_{j_pk_2} & \cdots & c_{j_pk_p} \end{vmatrix}.$$

Similarly

$$x_{K'}=\sum_{T'} e_{T'} C_{T'K'},$$

where $T=(t_1, t_2, \ldots, t_p)$ and T' are to be understood in the same way as K and K'. We now have

$$\begin{aligned} x_K\wedge x_{K'} &=\sum_{J,T}(e_J\wedge e_{T'})C_{JK}C_{T'K'} \\ &=\sum_J (e_J\wedge e_{J'})C_{JK}C_{J'K'} \end{aligned} \tag{5.3.6}$$

because $e_J \wedge e_{T'}=0$ unless $J=T$. Next, from (5.3.5) we obtain

$$\begin{aligned} x_K \wedge x_{K'} &= \varepsilon_K f(e_1) \wedge f(e_2) \wedge \cdots \wedge f(e_n) \\ &= \varepsilon_K \operatorname{Det}(C)(e_1 \wedge e_2 \wedge \cdots \wedge e_n), \end{aligned}$$

where ε_K is the sign of $(k_1, \ldots, k_p, k'_1, k'_2, \ldots)$ considered as a permutation of $(1, 2, 3, \ldots, n)$; that is to say ε_K is -1 raised to the power $(k_1-1)+(k_2-2)+\cdots+(k_p-p)$. We also have $e_J \wedge e_{J'}=\varepsilon_J(e_1 \wedge e_2 \wedge \cdots \wedge e_n)$. If we now substitute in (5.3.6) and remember that $e_1 \wedge e_2 \wedge \cdots \wedge e_n$ is actually a *base* of $E_n(F)$, we find that

$$\varepsilon_K \operatorname{Det}(C)=\sum_J \varepsilon_J C_{JK} C_{J'K'}.$$

Thus

$$\operatorname{Det}(C)=\sum_J \varepsilon_{JK} C_{JK} C_{J'K'} \tag{5.3.7}$$

where $\varepsilon_{JK}=\varepsilon_J\varepsilon_K$ is -1 raised to the power

$$(j_1+j_2+\cdots+j_p)+(k_1+k_2+\cdots+k_p).$$

This expression for $\operatorname{Det}(C)$ is known as *Laplace's expansion* of the determinant of the matrix C using the columns of the matrix that are indexed by $k_1, k_2, \ldots, k_p$.

We give one more illustration of the connection between the theory of exterior algebras and the theory of determinants. To this end consider free R-modules U, V and W of finite rank. We select bases for these, namely $u_1, u_2, \ldots, u_m$ for U, $v_1, v_2, \ldots, v_n$ for V and $w_1, w_2, \ldots, w_q$ for W. Let $\lambda: U \to V$ and $\mu: V \to W$ be R-linear mappings and let their matrices, with respect to the chosen bases, be $A=\|a_{ji}\|$ and $B=\|b_{kj}\|$ respectively. Thus A is an $n \times m$ matrix, B is a $q \times n$ matrix,

$$\lambda(u_i)=\sum_{j=1}^{n} v_j a_{ji},$$

and

$$\mu(v_j)=\sum_{k=1}^{q} w_k b_{kj}.$$

Furthermore the matrix C of $\mu \circ \lambda$ is given by $C=BA$.

After these preliminaries let $I=(i_1, i_2, \ldots, i_p)$ and $J=(j_1, j_2, \ldots, j_p)$ be sequences of p integers, where $1 \le i_1 < i_2 < \cdots < i_p \le m$ and $1 \le j_1 < j_2 < \cdots < j_p \le n$. Put $u_I=u_{i_1} \wedge u_{i_2} \wedge \cdots \wedge u_{i_p}$ and $v_J=v_{j_1} \wedge v_{j_2} \wedge \cdots \wedge v_{j_p}$. Then, by Lemma 1,

$$E_p(\lambda)(u_I)=\lambda(u_{i_1}) \wedge \lambda(u_{i_2}) \wedge \cdots \wedge \lambda(u_{i_p})=\sum_J v_J A_{JI}^{(p)},$$

where

$$A_{JI}^{(p)} = \begin{vmatrix} a_{j_1 i_1} & a_{j_1 i_2} & \cdots & a_{j_1 i_p} \\ a_{j_2 i_1} & a_{j_2 i_2} & \cdots & a_{j_2 i_p} \\ \vdots & \vdots & \vdots & \vdots \\ a_{j_p i_1} & a_{j_p i_2} & \cdots & a_{j_p i_p} \end{vmatrix}.$$

In a similar manner we can show that

$$E_p(\mu)(v_J) = \sum_K w_K B_{KJ}^{(p)},$$

where $K=(k_1, k_2, \ldots, k_p)$ is a sequence of integers satisfying $1 \leq k_1 < k_2 < \cdots < k_p \leq q$, $w_K = w_{k_1} \wedge w_{k_2} \wedge \cdots \wedge w_{k_p}$, and the definition of $B_{KJ}^{(p)}$ mimics that of $A_{JI}^{(p)}$. On the other hand

$$E_p(\mu \circ \lambda)(u_I) = \sum_K w_K C_{KI}^{(p)}$$

because $C = BA$ is the matrix of $\mu \circ \lambda$. But the w_K form a base for $E_p(W)$ and $E_p(\mu \circ \lambda) = E_p(\mu) \circ E_p(\lambda)$. Consequently

$$(BA)_{KI}^{(p)} = \sum_J B_{KJ}^{(p)} A_{JI}^{(p)}. \tag{5.3.8}$$

We mention briefly that if we fix p and regard the $A_{JI}^{(p)}$ as entries in a matrix $A^{(p)}$, then $A^{(p)}$ is known as the *p-th exterior power* of the matrix A. The relation (5.3.8) then yields $(BA)^{(p)} = B^{(p)} A^{(p)}$, which is the usual formula for the p-th exterior power of the product of two matrices.

5.4 The exterior algebra of a direct sum

Let $M_1, M_2, \ldots, M_q$ be R-modules and let $E(M_i)$ denote the exterior algebra of M_i. Then, using the construction set out in Section (3.4), we can form the *modified* tensor product $E(M_1) \underline{\otimes} E(M_2) \underline{\otimes} \cdots \underline{\otimes} E(M_q)$. This, of course, is an R-algebra and we know (Chapter 3, Theorem 7) that it is anticommutative.

Next there is an R-linear mapping

$$\begin{aligned} \phi\colon & M_1 \oplus M_2 \oplus \cdots \oplus M_q \\ & \to E(M_1) \underline{\otimes} E(M_2) \underline{\otimes} \cdots \underline{\otimes} E(M_q) \end{aligned} \tag{5.4.1}$$

given by

$$\phi(m_1, m_2, \ldots, m_q) = \sum_{i=1}^{q} 1 \otimes \cdots \otimes m_i \otimes \cdots \otimes 1. \tag{5.4.2}$$

(Naturally in $1 \otimes \cdots \otimes m_i \otimes \cdots \otimes 1$ it is understood that m_i occurs in the i-th position.) Note that $\phi(M_1 \otimes \cdots \otimes M_q)$ generates $E(M_1) \underline{\otimes} \cdots \underline{\otimes} E(M_q)$ as an R-algebra and $(\phi(m_1, m_2, \ldots, m_q))^2 = 0$ for $(m_1, m_2, \ldots, m_q)$ in $M_1 \oplus M_2 \oplus \cdots \oplus M_q$.

Theorem 8. *The modified tensor product*

$$E(M_1)\,\underline{\otimes}_R\,E(M_2)\,\underline{\otimes}_R\cdots\underline{\otimes}_R\,E(M_q), \tag{5.4.3}$$

together with the R-linear mapping ϕ defined in (5.4.2), *constitutes the exterior algebra of the direct sum $M_1\oplus M_2\oplus\cdots\oplus M_q$. Furthermore the grading which* (5.4.3) *possesses by virtue of being a modified tensor product is the same as its grading as the exterior algebra of $M_1\oplus M_2\oplus\cdots\oplus M_q$.*

Proof. For the moment let us leave on one side the assertions about gradings. For the rest, the isomorphisms

$$(M_1\oplus\cdots\oplus M_{q-1})\oplus M_q\approx M_1\oplus\cdots\oplus M_{q-1}\oplus M_q$$

and (Chapter 3, Theorem 6)

$$(E(M_1)\,\underline{\otimes}\cdots\underline{\otimes}\,E(M_{q-1}))\,\underline{\otimes}\,E(M_q)$$
$$\approx E(M_1)\,\underline{\otimes}\cdots\underline{\otimes}\,E(M_{q-1})\,\underline{\otimes}\,E(M_q)$$

show that the first part of the theorem will follow if we can establish it when $q=2$.

Suppose then that U and V are R-modules, put $M=U\oplus V$, and define the R-linear mapping $\phi\colon M\to E(U)\,\underline{\otimes}\,E(V)$ by $\phi(u,v)=u\otimes 1+1\otimes v$. We have to show that $E(U)\,\underline{\otimes}\,E(V)$ and ϕ constitute the exterior algebra of M. As a first step we note that, since $(\phi(u,v))^2=0$ for all $u\in U$ and $v\in V$, Theorem 2 shows that ϕ extends to a homomorphism $\lambda\colon E(M)\to E(U)\,\underline{\otimes}\,E(V)$ of R-algebras. Naturally $\lambda(u,v)=u\otimes 1+1\otimes v$.

Next the inclusion mappings $U\to M$ and $V\to M$ extend to homomorphisms $h\colon E(U)\to E(M)$ and $k\colon E(V)\to E(M)$ of graded R-algebras. Let us consider the mapping $E(U)\times E(V)\to E(M)$ in which (x,y) becomes $h(x)\wedge k(y)$. This is a bilinear mapping of R-modules. From this, and because $E(U)\otimes E(V)$ and $E(U)\,\underline{\otimes}\,E(V)$ coincide as R-modules, it follows that there exists an R-linear mapping

$$\lambda'\colon E(U)\,\underline{\otimes}\,E(V)\to E(M),$$

where $\lambda'(x\otimes y)=h(x)\wedge k(y)$ and $\lambda'(u\otimes 1+1\otimes v)=(u,v)$ for u in U and v in V.

Let $x\in E_r(U)$, $y\in E_s(V)$, $\xi\in E_\rho(U)$ and $\eta\in E_\sigma(V)$. By (3.4.5), the product of $x\otimes y$ and $\xi\otimes\eta$ in $E(U)\,\underline{\otimes}\,E(V)$ is $(-1)^{s\rho}(x\wedge\xi)\otimes(y\wedge\eta)$ and the image of this under λ' is

$$(-1)^{s\rho}h(x)\wedge h(\xi)\wedge k(y)\wedge k(\eta).$$

But h and k preserve degrees and $E(M)$ is anticommutative. Consequently the image in question is

$$(h(x)\wedge k(y))\wedge(h(\xi)\wedge k(\eta))=\lambda'(x\otimes y)\wedge\lambda'(\xi\otimes\eta).$$

It is now a simple matter to check that λ' is a homomorphism of R-algebras.

Consider $\lambda' \circ \lambda$. This is an algebra-homomorphism of $E(M)$ into itself and it induces the identity mapping on M. Consequently $\lambda' \circ \lambda$ is the identity mapping of $E(M)$. Next $\lambda' \circ \lambda$ is the identity mapping of $E(U) \underline{\otimes} E(V)$ because it induces the identity mapping on $\phi(M)$ and we know that, as an algebra, $E(U) \underline{\otimes} E(V)$ is generated by $\phi(M)$. Hence λ is an isomorphism of algebras and therefore $E(U) \underline{\otimes} E(V)$, together with ϕ, is the exterior algebra of M. As already observed, the first part of the theorem follows in full generality.

Finally, let us consider the two gradings on $E(M_1) \underline{\otimes} E(M_2) \underline{\otimes} \cdots \underline{\otimes} E(M_q)$ that are mentioned in the statement of the theorem. In both cases $\phi(M_1 \oplus \cdots \oplus M_q)$ is the module of elements of degree one. But $\phi(M_1 \oplus \cdots \oplus M_q)$ generates the algebra and therefore the two gradings must coincide.

5.5 Covariant extension of an exterior algebra

Let M be an R-module and $\omega: R \to S$ a homomorphism of commutative rings. Since we are here concerned with more than one commutative ring, we shall, for greater definiteness, use $E_R(M)$ to denote the exterior algebra of M. Of course $E_R(M)$ is a graded R-algebra, and therefore (see Section (3.6)) $E_R(M) \otimes_R S$ is an S-algebra graded by the S-modules $\{E_n(M) \otimes_R S\}_{n \in \mathbb{Z}}$. In particular $M \otimes_R S$ is the module of elements of degree one. It is easily verified that, as an S-algebra, $E_R(M) \otimes_R S$ is generated by $M \otimes_R S$ and that, in $E_R(M) \otimes_R S$, the square of an element of $M \otimes_R S$ is zero. It follows, from Lemma 3 of Chapter 3, that $E_R(M) \otimes_R S$ is anticommutative.

By Theorem 2, the inclusion mapping $M \otimes_R S \to E_R(M) \otimes_R S$ extends to a homomorphism

$$\lambda: E_S(M \otimes_R S) \to E_R(M) \otimes_R S$$

of S-algebras. (Here $E_S(M \otimes_R S)$ denotes the exterior algebra of the S-module $M \otimes_R S$.) Note that λ is degree-preserving.

Theorem 9. *The mapping*

$$\lambda: E_S(M \otimes_R S) \to E_R(M) \otimes_R S,$$

described above, is an isomorphism of graded S-algebras.

Proof. There is a close similarity between this result and Theorem 6 of Chapter 4. It will be found that it is a simple matter to adapt the arguments used to prove the earlier result, so the details have been omitted.

In view of Theorem 9 we may regard $E_R(M) \otimes_R S$ as the exterior algebra of the S-module $M \otimes_R S$. We shall therefore write

$$E_S(M \otimes_R S) = E_R(M) \otimes_R S \tag{5.5.1}$$

and, because λ is an isomorphism of *graded* algebras, we can add to this

$$E_n(M \otimes_R S) = E_n(M) \otimes_R S. \tag{5.5.2}$$

The latter relation holds for every integer n.

5.6 Skew derivations on an exterior algebra

The R-linear mappings of a module M into the ring R form the R-module $\mathrm{Hom}_R(M, R)$. We put

$$M^* = \mathrm{Hom}_R(M, R). \tag{5.6.1}$$

The members of M^* are, of course, the *linear forms* on M. M^* itself is known as the *dual* of M.

We recall that the structural homomorphism of the exterior algebra $E(M)$ maps R isomorphically onto $E_0(M)$. This fact will be used throughout Section (5.6) to identify the two so that in what follows we have

$$E_0(M) = R. \tag{5.6.2}$$

Now let Δ be a skew derivation on $E(M)$. Then Δ induces an R-homomorphism $E_1(M) \to E_0(M)$. But $E_1(M) = M$ and $E_0(M) = R$. Consequently Δ gives rise to a linear form on M. Furthermore, by Lemma 6 of Chapter 3, different skew derivations determine different linear forms.

Theorem 10. *Let f be a linear form on the R-module M. Then there is one and only one skew derivation on $E(M)$ which extends f.*

Proof. In view of what has already been said it will suffice to produce a skew derivation that agrees with f on $E_1(M) = M$. Suppose then that $p \geq 1$ is an integer and consider the mapping of $M \times M \times \cdots \times M$ (p factors) into $E_{p-1}(M)$ which takes $(m_1, m_2, \ldots, m_p)$ into

$$\sum_{i=1}^{p} (-1)^{i+1} f(m_i) m_1 \wedge \cdots \wedge \hat{m}_i \wedge \cdots \wedge m_p.$$

This mapping is multilinear and alternating. Consequently there is induced a homomorphism Δ_p: $E_p(M) \to E_{p-1}(M)$ of R-modules with the property that

$$\Delta_p(m_1 \wedge m_2 \wedge \cdots \wedge m_p)$$
$$= \sum_{i=1}^{p} (-1)^{i+1} f(m_i) m_1 \wedge \cdots \wedge \hat{m}_i \wedge \cdots \wedge m_p. \tag{5.6.3}$$

Note that when $p = 1$, Δ_p coincides with f.

Next, the various Δ_p ($p = 1, 2, 3, \ldots$) are the restrictions of a single R-endomorphism Δ: $E(M) \to E(M)$ that has degree -1; and now it only remains for us to verify that Δ is a skew derivation.

Let $x \in E_r(M)$ and $y \in E_s(M)$. We wish to show that $\Delta(x \wedge y)$ is equal to

$\Delta(x)\wedge y+(-1)^r x\wedge\Delta(y)$ and for this we may assume that $r\geq 1$ and $s\geq 1$. Indeed it is enough to establish the relation in question when $x=m_1\wedge m_2\wedge\cdots\wedge m_r$ and $y=\mu_1\wedge\mu_2\wedge\cdots\wedge\mu_s$, where the m_i and μ_j are elements of M. But then, by (5.6.3),

$$\begin{aligned}\Delta(x\wedge y)=&\sum_{i=1}^{r}(-1)^{i+1}f(m_i)m_1\wedge\cdots\wedge\hat{m}_i\wedge\cdots\wedge m_r\wedge\mu_1\wedge\cdots\wedge\mu_s\\&+\sum_{j=1}^{s}(-1)^{r+j+1}f(\mu_j)m_1\wedge\cdots\wedge m_r\wedge\mu_1\wedge\cdots\wedge\hat{\mu}_j\wedge\cdots\wedge\mu_s\\=&\,\Delta(x)\wedge y+(-1)^r x\wedge\Delta(y)\end{aligned}$$

and with this the proof is complete.

Suppose now that f is a linear form on M. By Theorem 10, it has a unique extension, Δ_f say, to a skew derivation on $E(M)$. We have seen that the skew derivations on $E(M)$ form an R-module and it is evident that

$$\Delta_{f_1+f_2}=\Delta_{f_1}+\Delta_{f_2} \tag{5.6.4}$$

and

$$\Delta_{rf}=r\Delta_f \tag{5.6.5}$$

for $f_1, f_2\in M^*$ and $r\in R$. Furthermore, by Lemma 6 of Chapter 3,

$$\Delta_f\circ\Delta_f=0. \tag{5.6.6}$$

Before we start to examine the consequences of these relations it will be convenient to broaden the basis of the discussion. To this end we note first that if

$$\gamma_0\colon M^*\times M\to R$$

is defined by $\gamma_0(f, m)=f(m)$, then γ_0 is bilinear and the mapping $M\to R$ in which m becomes $\gamma_0(f, m)$ is just f.

Suppose now that U is an R-module and that we have a bilinear mapping

$$\gamma\colon U\times M\to R, \tag{5.6.7}$$

i.e. suppose that we have a *bilinear form* on $U\times M$. If we fix u in U, then there is a linear form on M that maps m into $\gamma(u, m)$. This linear form will have a unique extension, Δ_u say, to a skew derivation on $E(M)$. Thus

$$\Delta_u(m)=\gamma(u, m) \tag{5.6.8}$$

for all $m\in M$. Furthermore (5.6.4), (5.6.5) and (5.6.6) generalize to give

$$\Delta_{u_1+u_2}=\Delta_{u_1}+\Delta_{u_2}, \tag{5.6.9}$$

$$\Delta_{ru}=r\Delta_u, \tag{5.6.10}$$

$$\Delta_u\circ\Delta_u=0, \tag{5.6.11}$$

where the notation is self-explanatory. It follows that there is an R-linear mapping

$$U \to \mathrm{End}_R(E(M)) \tag{5.6.12}$$

in which u has image Δ_u. Note that if $q \geq 1$ and $m_1, m_2, \ldots, m_q$ are in M, then, by (3.7.4) and (5.6.8),

$$\Delta_u(m_1 \wedge m_2 \wedge \cdots \wedge m_q) = \sum_{i=1}^{q} (-1)^{i+1} \gamma(u, m_i) m_1 \wedge \cdots \wedge \hat{m}_i \wedge \cdots \wedge m_q \tag{5.6.13}$$

for all u in U. But $\Delta_u \circ \Delta_u = 0$. Accordingly (5.6.12) extends to a homomorphism

$$\Gamma: E(U) \to \mathrm{End}_R(E(M)) \tag{5.6.14}$$

of R-algebras. Note that if $u_1, u_2, \ldots, u_p$ belong to U, then

$$\Gamma(u_1 \wedge u_2 \wedge \cdots \wedge u_p) = \Gamma(u_1) \circ \Gamma(u_2) \circ \cdots \circ \Gamma(u_p)$$

and therefore

$$\Gamma(u_1 \wedge u_2 \wedge \cdots \wedge u_p) = \Delta_{u_1} \circ \Delta_{u_2} \circ \cdots \circ \Delta_{u_p}. \tag{5.6.15}$$

This shows that when $\Gamma(u_1 \wedge u_2 \wedge \cdots \wedge u_p)$ operates on a homogeneous element of $E(M)$ it lowers its degree by p. We also see that for $p \geq 2$

$$\Gamma(u_1 \wedge u_2 \wedge \cdots \wedge u_p) = \Delta_{u_1} \circ \Gamma(u_2 \wedge u_3 \wedge \cdots \wedge u_p). \tag{5.6.16}$$

We now wish to make explicit the way in which $\Gamma(u_1 \wedge u_2 \wedge \cdots \wedge u_p)$ operates on $E(M)$ and for this it is convenient to introduce some temporary notation.

Suppose that $1 \leq p \leq q$. We shall use $T = (t_1, t_2, \ldots, t_p)$ to denote a sequence of p integers satisfying $1 \leq t_1 < t_2 < \cdots < t_p \leq q$; and if $m_1, m_2, \ldots, m_q$ belong to M, then $(m_1 \wedge m_2 \wedge \cdots \wedge m_q)_T$ will denote the result of striking out from $m_1 \wedge m_2 \wedge \cdots \wedge m_q$ the terms $m_{t_1}, m_{t_2}, \ldots, m_{t_p}$. Thus

$$(m_1 \wedge m_2 \wedge \cdots \wedge m_q)_T = m_1 \wedge \cdots \wedge \hat{m}_{t_1} \wedge \cdots \wedge \hat{m}_{t_p} \wedge \cdots \wedge m_q. \tag{5.6.17}$$

(If $q = p$, then $(m_1 \wedge m_2 \wedge \cdots \wedge m_q)_T$ is understood to be the identity element of $E(M)$.) Finally we put

$$D_T = \begin{vmatrix} \gamma(u_1, m_{t_1}) & \gamma(u_1, m_{t_2}) & \cdots & \gamma(u_1, m_{t_p}) \\ \gamma(u_2, m_{t_1}) & \gamma(u_2, m_{t_2}) & \cdots & \gamma(u_2, m_{t_p}) \\ \vdots & \vdots & \vdots\vdots & \vdots \\ \gamma(u_p, m_{t_1}) & \gamma(u_p, m_{t_2}) & \cdots & \gamma(u_p, m_{t_p}) \end{vmatrix} \tag{5.6.18}$$

and

$$|T| = t_1 + t_2 + \cdots + t_p.$$

Lemma 2. *Suppose that $1 \leq p \leq q$ and let $u_1, u_2, \ldots, u_p$ belong to U and $m_1, m_2, \ldots, m_q$ to M. Then, with the above notation,*

$$(\Gamma(u_1 \wedge u_2 \wedge \cdots \wedge u_p))(m_1 \wedge m_2 \wedge \cdots \wedge m_q)$$
$$= (-1)^p \sum_T (-1)^{|T|} D_T (m_1 \wedge m_2 \wedge \cdots \wedge m_q)_T. \tag{5.6.19}$$

Proof. We argue by induction on p. If $u \in U$, then

$$(\Gamma(u))(m_1 \wedge m_2 \wedge \cdots \wedge m_q) = \Delta_u(m_1 \wedge m_2 \wedge \cdots \wedge m_q)$$
$$= \sum_{i=1}^{q} (-1)^{i+1} \gamma(u, m_i) m_1 \wedge \cdots \wedge \hat{m}_i \wedge \cdots \wedge m_q$$

by (5.6.13). This shows that (5.6.19) holds when $p=1$.

We now assume that $p>1$ and that the relation in question has been established for all smaller values of the inductive variable. Then, by the inductive hypothesis,

$$(\Gamma(u_2 \wedge u_3 \wedge \cdots \wedge u_p))(m_1 \wedge m_2 \wedge \cdots \wedge m_q)$$
$$= (-1)^{p-1} \sum_K (-1)^{|K|} D'_K (m_1 \wedge m_2 \wedge \cdots \wedge m_q)_K. \tag{5.6.20}$$

Here the sequence $K=(k_2, k_3, \ldots, k_p)$ is required to satisfy $1 \le k_2 < k_3 < \cdots < k_p \le q$ and

$$D'_K = \begin{vmatrix} \gamma(u_2, m_{k_2}) & \gamma(u_2, m_{k_3}) & \cdots & \gamma(u_2, m_{k_p}) \\ \gamma(u_3, m_{k_2}) & \gamma(u_3, m_{k_3}) & \cdots & \gamma(u_3, m_{k_p}) \\ \vdots & \vdots & \vdots\vdots\vdots & \vdots \\ \gamma(u_p, m_{k_2}) & \gamma(u_p, m_{k_3}) & \cdots & \gamma(u_p, m_{k_p}) \end{vmatrix}.$$

Next, if we operate with Δ_{u_1} on (5.6.20) and use (5.6.16) we find that

$$(\Gamma(u_1 \wedge u_2 \wedge \cdots \wedge u_p))(m_1 \wedge m_2 \wedge \cdots \wedge m_q)$$
$$= (-1)^{p-1} \sum_K (-1)^{|K|} D'_K \Delta_{u_1}((m_1 \wedge m_2 \wedge \cdots \wedge m_q)_K). \tag{5.6.21}$$

We must now identify the coefficient of $(m_1 \wedge m_2 \wedge \cdots \wedge m_q)_T$ when we expand the right-hand side of (5.6.21). To this end, for $1 \le v \le p$ we put $K_v = (t_1, \ldots, \hat{t}_v, \ldots, t_p)$. Then when $\Delta_{u_1}((m_1 \wedge m_2 \wedge \cdots \wedge m_q)_{K_v})$ is developed it provides the term

$$\gamma(u_1, m_{t_v})(-1)^{t_v + v}(m_1 \wedge m_2 \wedge \cdots \wedge m_q)_T.$$

We also have $|K_v| + t_v = |T|$. Hence the coefficient of $(m_1 \wedge m_2 \wedge \cdots \wedge m_q)_T$ in the expansion of the right-hand side of (5.6.21) is

$$(-1)^{p-1} \sum_{v=1}^{p} (-1)^{|T|+v} \gamma(u_1, m_{t_v}) D'_{K_v}.$$

and therefore to complete the proof it is enough to show that

$$D_T = \sum_{v=1}^{p} (-1)^{v+1} \gamma(u_1, m_{t_v}) D'_{K_v}.$$

However, this follows on expanding the determinant (5.6.18) by means of its first row.

Corollary. *Suppose that $p \geq 1$. Let $u_1, u_2, \ldots, u_p$ belong to U and $m_1, m_2, \ldots, m_p$ to M. Then*

$$(\Delta_{u_1} \circ \Delta_{u_2} \circ \cdots \circ \Delta_{u_p})(m_1 \wedge m_2 \wedge \cdots \wedge m_p) = (-1)^{p(p+3)/2} \begin{vmatrix} \gamma(u_1, m_1) & \gamma(u_1, m_2) & \cdots & \gamma(u_1, m_p) \\ \gamma(u_2, m_1) & \gamma(u_2, m_2) & \cdots & \gamma(u_2, m_p) \\ \vdots & \vdots & \vdots\vdots\vdots & \vdots \\ \gamma(u_p, m_1) & \gamma(u_p, m_2) & \cdots & \gamma(u_p, m_p) \end{vmatrix}.$$

Proof. If we take account of (5.6.15) this is simply what (5.6.19) becomes when $p=q$.

5.7 Pfaffians

Let M be an R-module and let

$$\gamma \colon M \times M \to R \tag{5.7.1}$$

be an *alternating* bilinear mapping of $M \times M$ into the ring R (i.e. γ is an alternating bilinear *form*). Furthermore for $m \in M$ let Δ_m denote the skew derivation on $E(M)$ which satisfies

$$\Delta_m(\mu) = \gamma(m, \mu) \tag{5.7.2}$$

when μ belongs to M. (Note that we are continuing to identify $E_0(M)$ with R.) Then if $m_1, m_2 \in M$ and $r \in R$, we have $\Delta_{m_1+m_2} = \Delta_{m_1} + \Delta_{m_2}$ and $\Delta_{rm} = r\Delta_m$. Next for $m \in M$ we define an endomorphism

$$L_m \colon E(M) \to E(M) \tag{5.7.3}$$

by

$$L_m(x) = m \wedge x. \tag{5.7.4}$$

Note that $L_{m_1+m_2} = L_{m_1} + L_{m_2}$, $L_{rm} = rL_m$ and, whereas Δ_m has degree -1, the degree of L_m is $+1$. We now put

$$\Lambda_m = L_m + \Delta_m. \tag{5.7.5}$$

Of course Λ_m is also an endomorphism of the R-module $E(M)$.

Lemma 3. *The mapping*

$$M \to \mathrm{End}_R(E(M)) \tag{5.7.6}$$

in which m of M becomes Λ_m is R-linear. Furthermore $\Lambda_m \circ \Lambda_m = 0$ for all $m \in M$.

Proof. The first assertion is clear. Now let $x \in E(M)$. Then $(L_m \circ L_m)(x) = m \wedge m \wedge x = 0$, so that $L_m \circ L_m = 0$. We also have $\Delta_m \circ \Delta_m = 0$ by (5.6.11). Next

$$(\Delta_m \circ L_m)(x) = \Delta_m(m \wedge x) = \gamma(m, m)x - m \wedge \Delta_m(x).$$

But $\gamma(m, m)=0$ because γ is an *alternating* bilinear form. Consequently

$$(\Delta_m \circ L_m)(x) = -m \wedge \Delta_m(x) = -(L_m \circ \Delta_m)(x)$$

and therefore $\Delta_m \circ L_m + L_m \circ \Delta_m = 0$. The lemma follows.

By Theorem 2, the mapping (5.7.6) extends to a homomorphism

$$E(M) \to \mathrm{End}_R(E(M)) \tag{5.7.7}$$

of R-algebras. Let us denote the image of an element x of $E(M)$ by Λ_x. Then

$$\Lambda_{m_1 \wedge m_2 \wedge \cdots \wedge m_p} = (L_{m_1} + \Delta_{m_1}) \circ (L_{m_2} + \Delta_{m_2}) \circ \cdots \circ (L_{m_p} + \Delta_{m_p}) \tag{5.7.8}$$

whenever $m_1, m_2, \ldots, m_p$ belong to M.

We next define

$$\Omega: E(M) \to E(M) \tag{5.7.9}$$

by

$$\Omega(x) = \Lambda_x(1_{E(M)}). \tag{5.7.10}$$

Clearly Ω is an endomorphism of the R-module $E(M)$ and $\Omega(1_{E(M)}) = 1_{E(M)}$. It follows that Ω *induces the identity mapping on* $E_0(M)$. Furthermore if $m \in M$ and $u \in E(M)$, then

$$\Omega(m \wedge u) = (\Lambda_m \circ \Lambda_u)(1_{E(M)}) = \Lambda_m(\Omega(u))$$

and hence

$$\Omega(m \wedge u) = m \wedge \Omega(u) + \Delta_m(\Omega(u)). \tag{5.7.11}$$

Theorem 11. *Let Ω be defined as above. Then for every skew derivation Δ on $E(M)$ we have $\Omega \circ \Delta = \Delta \circ \Omega$.*

Proof. Suppose that $n \geq 0$. It will suffice to prove that *if $x \in E_n(M)$ and Δ is a skew derivation, then $\Omega(\Delta(x)) = \Delta(\Omega(x))$*. This will be established by induction on n.

First we note that if $x \in E_0(M)$, then $\Omega(\Delta(x))$ and $\Delta(\Omega(x))$ are both zero. Thus all is well when $n=0$.

From here on we assume that $n \geq 1$ and that the assertion in question has been established for all smaller non-negative values of the inductive variable. Accordingly we have $\Omega(\Delta'(z)) = \Delta'(\Omega(z))$ if Δ' is a skew derivation and $z \in E_{n-1}(M)$.

Now let $x \in E_n(M)$ and let Δ be a skew derivation. We wish to show that $\Omega(\Delta(x))$ and $\Delta(\Omega(x))$ are the same and for this step we may assume that $x = m \wedge y$, where $m \in M$ and $y \in E_{n-1}(M)$. Now $\Delta(m)$ belongs to $E_0(M) = R$. Consequently

$$\begin{aligned}\Omega(\Delta(x)) &= \Omega(\Delta(m)y - m \wedge \Delta(y)) \\ &= \Delta(m)\Omega(y) - m \wedge \Omega(\Delta(y)) - (\Delta_m \circ \Omega)(\Delta(y))\end{aligned}$$

by (5.7.11). On the other hand, again by (5.7.11),

$$\begin{aligned}\Delta(\Omega(x)) &= \Delta(\Omega(m \wedge y)) \\ &= \Delta(m \wedge \Omega(y) + \Delta_m(\Omega(y))) \\ &= \Delta(m)\Omega(y) - m \wedge \Delta(\Omega(y)) + (\Delta \circ \Delta_m)(\Omega(y)).\end{aligned}$$

But $\Delta(\Omega(y)) = \Omega(\Delta(y))$ by the inductive hypothesis and $\Delta \circ \Delta_m = -\Delta_m \circ \Delta$ by Lemma 6 of Chapter 3. Accordingly

$$\begin{aligned}\Delta(\Omega(x)) &= \Delta(m)\Omega(y) - m \wedge \Omega(\Delta(y)) - (\Delta_m \circ \Omega)(\Delta(y)) \\ &= \Omega(\Delta(x)).\end{aligned}$$

The lemma follows.

We now introduce a second alternating bilinear form

$$\bar{\gamma}: M \times M \to R \tag{5.7.12}$$

by putting

$$\bar{\gamma}(m_1, m_2) = -\gamma(m_1, m_2) \tag{5.7.13}$$

and using this we define $\bar{\Delta}_m$, $\bar{\Lambda}_x$ and $\bar{\Omega}$ just as we defined Δ_m, Λ_x and Ω using γ. By (5.7.13)

$$\bar{\Delta}_m = -\Delta_m. \tag{5.7.14}$$

Of course, like Ω, $\bar{\Omega}$ is an endomorphism of $E(M)$ and it commutes with all skew derivations. Suppose that $m \in M$ and $u \in E(M)$. Then, in view of (5.7.14) and (5.7.11), we have

$$\bar{\Omega}(m \wedge u) = m \wedge \bar{\Omega}(u) - \Delta_m(\bar{\Omega}(u)). \tag{5.7.15}$$

We also have, this time from (5.7.8),

$$\bar{\Lambda}_{m_1 \wedge m_2 \wedge \cdots \wedge m_p} = (L_{m_1} - \Delta_{m_1}) \circ (L_{m_2} - \Delta_{m_2}) \circ \cdots \circ (L_{m_p} - \Delta_{m_p}) \tag{5.7.16}$$

for $m_1, m_2, \ldots, m_p$ in M.

Theorem 12. *Ω and $\bar{\Omega}$ are inverse automorphisms of the R-module $E(M)$.*

Proof. We shall establish the following statement by induction on n: *for all $n \geq 0$, if $x \in E_n(M)$, then $\bar{\Omega}(\Omega(x)) = x$ and $\Omega(\bar{\Omega}(x)) = x$.* This will clearly suffice.

The statement is true when $n = 0$ because both Ω and $\bar{\Omega}$ induce the identity mapping on $E_0(M)$. From here on we shall suppose that $n \geq 1$ and make the natural inductive hypothesis.

Let $x \in E_n(M)$. In showing that $\bar{\Omega}(\Omega(x)) = x$ we may suppose that $x = m \wedge y$, where $m \in M$ and $y \in E_{n-1}(M)$. But then, by (5.7.11) and (5.7.15),

$$\begin{aligned}\bar{\Omega}(\Omega(x)) &= \bar{\Omega}\{m \wedge \Omega(y) + \Delta_m(\Omega(y))\} \\ &= m \wedge \bar{\Omega}(\Omega(y)) - \Delta_m(\bar{\Omega}(\Omega(y))) + \bar{\Omega}(\Delta_m(\Omega(y))) \\ &= m \wedge y - \Delta_m(y) + \Delta_m(y) \\ &= m \wedge y \\ &= x\end{aligned}$$

because $\bar{\Omega}(\Omega(y))=y$ by the inductive hypothesis and because $\bar{\Omega}$ commutes with Δ_m. Since $\Omega(\bar{\Omega}(x))=x$ for entirely similar reasons, the proof is complete.

Theorem 13. *Let $m_1, m_2, \ldots, m_p$ $(p \geq 1)$ belong to M. Then*

$$\begin{vmatrix} \gamma(m_1, m_1) & \gamma(m_1, m_2) & \cdots & \gamma(m_1, m_p) \\ \gamma(m_2, m_1) & \gamma(m_2, m_2) & \cdots & \gamma(m_2, m_p) \\ \vdots & \vdots & & \vdots \\ \gamma(m_p, m_1) & \gamma(m_p, m_2) & \cdots & \gamma(m_p, m_p) \end{vmatrix} = \omega^2, \tag{5.7.17}$$

where ω is the component of degree zero of the result of applying

$$(L_{m_1}+\Delta_{m_1})\circ(L_{m_2}+\Delta_{m_2})\circ\cdots\circ(L_{m_p}+\Delta_{m_p})$$

to $1_{E(M)}$. If p is odd, then the determinant and ω are both zero.

Remark. Theorem 13 is of interest not only because it shows that the determinant is a perfect square but also because it identifies its square root.

Proof. For u in $E(M)$ let $\pi_0(u)$ denote its homogeneous component of degree zero. Let $x=m_1 \wedge m_2 \wedge \cdots \wedge m_p$. Then the corollary to Lemma 2 shows that the determinant in (5.7.17) is equal to

$$\begin{aligned}(-1)^{p(p+3)/2}&(\Delta_{m_1}\circ\Delta_{m_2}\circ\cdots\circ\Delta_{m_p})(x) \\ &=(-1)^{p(p+3)/2}\pi_0\{((L_{m_1}+\Delta_{m_1})\circ\cdots\circ(L_{m_p}+\Delta_{m_p}))(x)\} \\ &=(-1)^{p(p+3)/2}\pi_0(\Lambda_x(x)).\end{aligned}$$

Put $\bar{x}=\bar{\Omega}(x)$. By Theorem 12, $x=\Omega(\bar{x})$ and therefore

$$\begin{aligned}\Lambda_x(x)&=\Lambda_x(\Omega(\bar{x}))=\Lambda_x(\Lambda_{\bar{x}}(1_{E(M)})) \\ &=(\Lambda_x\circ\Lambda_{\bar{x}})(1_{E(M)}) \\ &=\Lambda_{x\wedge\bar{x}}(1_{E(M)})=\Omega(x\wedge\bar{x}).\end{aligned}$$

But $\bar{x}=\bar{\Omega}(x)$ is the result of applying

$$(L_{m_1}-\Delta_{m_1})\circ(L_{m_2}-\Delta_{m_2})\circ\cdots\circ(L_{m_p}-\Delta_{m_p})$$

to $1_{E(M)}$ (see (5.7.16)) so that $\bar{x}$ belongs to the R-module spanned by all products $m_{i_1}\wedge m_{i_2}\wedge\cdots\wedge m_{i_h}$. However, $x\wedge m_{i_1}\wedge m_{i_2}\wedge\cdots\wedge m_{i_h}=0$ if $h>0$ and thus we see that $x\wedge\bar{x}=\pi_0(\bar{x})x$. Consequently $\Lambda_x(x)=\pi_0(\bar{x})\Omega(x)$ and therefore the determinant in (5.7.17) is equal to $(-1)^{p(p+3)/2}\pi_0(\bar{x})\pi_0(\Omega(x))$. On the other hand $\omega=\pi_0(\Omega(x))$ so that it follows that the relation (5.7.17) will be established if we show that

$$(-1)^{p(p+3)/2}\pi_0(\bar{x})=\pi_0(\Omega(x)). \tag{5.7.18}$$

Next, by (5.7.8),

$$\begin{aligned}\Lambda_x&=(L_{m_1}+\Delta_{m_1})\circ(L_{m_2}+\Delta_{m_2})\circ\cdots\circ(L_{m_p}+\Delta_{m_p}) \\ &=\sum_{s=0}^{p}\Lambda_{p-2s,x},\end{aligned}$$

where $\Lambda_{p-2s,x}$ is an endomorphism of $E(M)$ of degree $p-2s$, and now we may conclude, from (5.7.16), that

$$\bar{\Lambda}_x = \sum_{s=0}^{p} (-1)^s \Lambda_{p-2s,x}.$$

Accordingly

$$\Omega(x) = \Lambda_x(1_{E(M)}) = \sum_{s=0}^{p} \Lambda_{p-2s,x}(1_{E(M)})$$

and

$$\bar{x} = \bar{\Lambda}_x(1_{E(M)}) = \sum_{s=0}^{p} (-1)^s \Lambda_{p-2s,x}(1_{E(M)}).$$

Thus if p is *odd*, then both $\pi_0(\bar{x})=0$ and $\pi_0(\Omega(x))=0$ and therefore, in this case, both the determinant and ω are zero. On the other hand, if $p=2k$ then

$$\pi_0(\Omega(x)) = \Lambda_{0,x}(1_{E(M)}) = (-1)^k \pi_0(\bar{x}).$$

But $(-1)^{p(p+3)/2} = (-1)^k$ and therefore

$$\pi_0(\Omega(x)) = (-1)^{p(p+3)/2} \pi_0(\bar{x})$$

as required.

We shall now extract from Theorem 13 some information which is directly relevant to the theory of matrices. To this end let

$$A = \begin{bmatrix} a_{11} & a_{12} & \cdots & a_{1p} \\ a_{21} & a_{22} & \cdots & a_{2p} \\ \vdots & \vdots & \vdots\vdots\vdots & \vdots \\ a_{p1} & a_{p2} & \cdots & a_{pp} \end{bmatrix}$$

be a $p \times p$ matrix with entries in R. We call A an *alternating* matrix provided that (i) $a_{ii}=0$ for all i, and (ii) $a_{ji} = -a_{ij}$ for all i and j.

Suppose then that A is an alternating matrix. We can construct a free R-module M with a base $e_1, e_2, \ldots, e_p$ of p elements and afterwards an alternating bilinear form $\gamma: M \times M \to R$ with $\gamma(e_i, e_j) = a_{ij}$. This enables us to introduce the *Pfaffian* of A, which will be denoted by $\mathrm{Pf}(A)$. In fact the Pfaffian is defined by

$$\mathrm{Pf}(A) = \pi_0\{((L_{e_1} + \Delta_{e_1}) \circ (L_{e_2} + \Delta_{e_2}) \circ \cdots \circ (L_{e_n} + \Delta_{e_n})(1_{E(M)})\},$$

where $\pi_0: E(M) \to E_0(M)$ is the natural projection. It follows (Theorem 13) that

$$\mathrm{Det}(A) = (\mathrm{Pf}(A))^2 \qquad (5.7.19)$$

and *when p is odd* $\mathrm{Pf}(A)$ *is zero.*

By way of example consider a general 4×4 alternating matrix

$$A=\begin{bmatrix} 0 & a_{12} & a_{13} & a_{14} \\ -a_{12} & 0 & a_{23} & a_{24} \\ -a_{13} & -a_{23} & 0 & a_{34} \\ -a_{14} & -a_{24} & -a_{34} & 0 \end{bmatrix}.$$

In this case $\mathrm{Pf}(A)$ is the component of degree zero of

$$\{(L_{e_1}+\Delta_{e_1})\circ(L_{e_2}+\Delta_{e_2})\circ(L_{e_3}+\Delta_{e_3})\circ(L_{e_4}+\Delta_{e_4})\}(1_{E(M)}).$$

But L_{e_i} has degree $+1$ and Δ_{e_i} has degree -1. Consequently

$$\begin{aligned}\mathrm{Pf}(A)&=(\Delta_{e_1}\circ\Delta_{e_2}\circ L_{e_3}\circ L_{e_4})(1_{E(M)})+(\Delta_{e_1}\circ L_{e_2}\circ\Delta_{e_3}\circ L_{e_4})(1_{E(M)})\\ &=\Delta_{e_1}(\Delta_{e_2}(e_3\wedge e_4))+\Delta_{e_1}(L_{e_2}(a_{34}))\\ &=a_{23}a_{14}-a_{24}a_{13}+a_{12}a_{34}.\end{aligned}$$

Thus the relation $\mathrm{Det}(A)=(\mathrm{Pf}(A))^2$ becomes

$$\begin{vmatrix} 0 & a_{12} & a_{13} & a_{14} \\ -a_{12} & 0 & a_{23} & a_{24} \\ -a_{13} & -a_{23} & 0 & a_{34} \\ -a_{14} & -a_{24} & -a_{34} & 0 \end{vmatrix}=(a_{12}a_{34}-a_{13}a_{24}+a_{14}a_{23})^2$$

as may be verified directly.

5.8 Comments and exercises

The theory of exterior algebras is of particular interest because of its connection with the theory of determinants, and indeed most of what will be said here arises out of this connection. However, there are two comments on the general theory that come to mind. These will be dealt with first.

As usual, the examples for which solutions are provided are marked with an asterisk. If M is an R-module, then $E(M)$ – or $E_R(M)$ when it is desired to remind the reader of the ground ring in question – always denotes its exterior algebra.

First suppose that I is a proper ideal of R and that M is an R-module. Then $IE(M)$ is a homogeneous two-sided ideal of $E(M)$ and therefore $E(M)/IE(M)$ inherits the structure of a graded R-algebra. But I annihilates this algebra, so $E(M)/IE(M)$ can also be regarded as an R/I-algebra.

Next, the natural mapping $E(M)\to E(M)/IE(M)$ induces a homomorphism of $E_1(M)=M$ into $E(M)/IE(M)$ which vanishes on IM. Thus there is induced a homomorphism

$$M/IM\to E(M)/IE(M). \tag{5.8.1}$$

To be precise this is a homomorphism both of R-modules and of R/I-modules. Furthermore the image of every element of M/IM has its square equal to zero. It follows that (5.8.1) can be extended to a homomorphism

$$\phi: E_{R/I}(M/IM) \to E(M)/IE(M) \tag{5.8.2}$$

of R/I-algebras.

Exercise 1. *Show that the homomorphism* (5.8.2) *is an isomorphism of graded* R/I*-algebras.*

There is a solution to this exercise which is very similar to the solution provided for Exercise 5 of Chapter 4. We shall therefore leave the reader to make the minor adjustments that are required. Note that this result enables us to write

$$E_{R/I}(M/IM) = E_R(M)/IE_R(M). \tag{5.8.3}$$

Our next comment has to do with generalized skew derivations. We already know, from Theorem 10, that, for any R-module M, $E(M)$ is fully endowed with skew derivations. The next exercise extends this result.

Exercise 2. *Let M be an R-module, let i be a given integer, and let $f: M \to E_{i+1}(M)$ be an R-linear mapping. Show that there is one and only one generalized skew derivation of degree i, on $E(M)$, that extends f.*

This time we can adapt the solution to Exercise 6 of Chapter 4. The only point to be noted is that if $p \geq 1$, then the mapping of the p-fold product $M \times M \times \cdots \times M$ into $E_{i+p}(M)$ which takes $(m_1, m_2, \ldots, m_p)$ into

$$\sum_{\mu=1}^{p} (-1)^{(\mu+1)i} m_1 \wedge \cdots \wedge f(m_\mu) \wedge \cdots \wedge m_p$$

is an *alternating* multilinear mapping. This is because $E(M)$ is an anticommutative algebra (see Theorem 3).

Our next comments have to do with Laplace's expansion of a determinant and we begin by recalling what has been established already. Suppose that

$$C = \begin{bmatrix} c_{11} & c_{12} & \cdots & c_{1n} \\ c_{21} & c_{22} & \cdots & c_{2n} \\ \vdots & \vdots & \vdots & \vdots \\ c_{n1} & c_{n2} & \cdots & c_{nn} \end{bmatrix} \tag{5.8.4}$$

is an $n \times n$ matrix with entries in R and let p be an integer satisfying $1 < p < n$. In what follows $J = (j_1, j_2, \ldots, j_p)$ respectively $K = (k_1, k_2, \ldots, k_p)$ will denote a sequence of p integers with $1 \leq j_1 < \cdots < j_p \leq n$ respectively $1 \leq k_1 < \cdots < k_p \leq n$, and as before we put

$$C_{JK} = \begin{bmatrix} c_{j_1k_1} & c_{j_1k_2} & \cdots & c_{j_1k_p} \\ c_{j_2k_1} & c_{j_2k_2} & \cdots & c_{j_2k_p} \\ \vdots & \vdots & \vdots & \vdots \\ c_{j_pk_1} & c_{j_pk_2} & \cdots & c_{j_pk_p} \end{bmatrix}. \tag{5.8.5}$$

Again J' respectively K' will be used to denote the increasing sequence that is obtained by striking out from $(1, 2, \ldots, n)$ the terms that belong to J respectively K. If now we put

$$|J| = j_1 + j_2 + \cdots + j_p, \tag{5.8.6}$$

then (5.3.7) can be rewritten as

$$\operatorname{Det}(C) = \sum_J (-1)^{|J|+|K|} C_{JK} C_{J'K'}. \tag{5.8.7}$$

This, as we remarked at the time, is Laplace's expansion of the determinant of C using columns $k_1, k_2, \ldots, k_p$. Now there is a companion formula, namely

$$\operatorname{Det}(C) = \sum_K (-1)^{|J|+|K|} C_{JK} C_{J'K'}, \tag{5.8.8}$$

which is Laplace's expansion using rows $j_1, j_2, \ldots, j_p$. This can be derived by applying (5.8.7) to the transpose of C.

The following observation will serve to introduce the next exercise. Consider the 2×4 matrix

$$\begin{bmatrix} x_1 & x_2 & x_3 & x_4 \\ y_1 & y_2 & y_3 & y_4 \end{bmatrix}.$$

There is an identity that connects its 2×2 subdeterminants which takes the form

$$\begin{vmatrix} x_1 & x_2 \\ y_1 & y_2 \end{vmatrix} \cdot \begin{vmatrix} x_3 & x_4 \\ y_3 & y_4 \end{vmatrix} - \begin{vmatrix} x_1 & x_3 \\ y_1 & y_3 \end{vmatrix} \cdot \begin{vmatrix} x_2 & x_4 \\ y_2 & y_4 \end{vmatrix}$$

$$+ \begin{vmatrix} x_1 & x_4 \\ y_1 & y_4 \end{vmatrix} \cdot \begin{vmatrix} x_2 & x_3 \\ y_2 & y_3 \end{vmatrix} = 0. \tag{5.8.9}$$

Using Laplace's expansion, (5.8.9) can be generalized as follows.

Exercise 3*. *Let $A = ||a_{ij}||$ be an $n \times 2n$ matrix with entries in R, let $I = (i_1, i_2, \ldots, i_n)$ be a typical sequence of n integers satisfying $1 \le i_1 < \cdots < i_n \le 2n$, and let I' be the sequence obtained from $(1, 2, \ldots, 2n)$ by deleting the terms in I. If*

$$D_I = \begin{vmatrix} a_{1i_1} & a_{1i_2} & \cdots & a_{1i_n} \\ a_{2i_1} & a_{2i_2} & \cdots & a_{2i_n} \\ \vdots & \vdots & \vdots\vdots\vdots & \vdots \\ a_{ni_1} & a_{ni_2} & \cdots & a_{ni_n} \end{vmatrix}$$

show that

$$\sum_I \operatorname{sgn}(I, I') D_I D_{I'} = 0. \tag{5.8.10}$$

Show also that if k (where $1 \le k \le 2n$) is fixed and n is even, then

$$\sum_{k \in I} \operatorname{sgn}(I, I') D_I D_{I'} = 0. \tag{5.8.11}$$

Here it is to be understood that the summation is over the different sequences I, and that if $I' = (i'_1, i'_2, \ldots, i'_n)$, then by $\operatorname{sgn}(I, I')$ is meant the sign of $(i_1, i_2, \ldots, i'_1, i'_2, \ldots)$ considered as a permutation of $(1, 2, \ldots, 2n)$. Of course (5.8.9) is what (5.8.11) becomes when $n=2$.

We return to (5.8.8). Let $S = (s_1, s_2, \ldots, s_p)$, where $1 \le s_1 < \cdots < s_p \le n$, and let S' (the corresponding residual sequence) be $(s'_1, s'_2, \ldots, s'_{n-p})$. Further, let an $n \times n$ matrix Q be formed as follows: the first p rows of Q are to be rows $j_1, j_2, \ldots, j_p$ of the matrix C and the last $n-p$ rows of Q are to be rows $s'_1, s'_2, \ldots, s'_{n-p}$ of C. Evidently, if $J = (j_1, j_2, \ldots, j_p)$ and S are different, then the determinant of Q is zero whereas if $J = S$, then

$$\operatorname{Det}(Q) = \operatorname{sgn}(J, J') \operatorname{Det}(C)$$

and therefore

$$(-1)^{1+2+\cdots+p} \operatorname{Det}(Q) = (-1)^{|S|} \operatorname{Det}(C).$$

On the other hand, for arbitrary J and S

$$\operatorname{Det}(Q) = \sum_K (-1)^{1+2+\cdots+p} (-1)^{|K|} C_{JK} C_{S'K'}$$

as may be seen by employing Laplace's expansion on the first p rows of Q. It follows that

$$\sum_K (-1)^{|S|+|K|} C_{JK} C_{S'K'}$$

is zero if $J \neq S$ and it has the value $\operatorname{Det}(C)$ if $J = S$. Hence if we put

$$\Gamma_{KS} = (-1)^{|K|+|S|} C_{S'K'} \tag{5.8.12}$$

then we obtain

$$\sum_K C_{JK} \Gamma_{KS} = \begin{cases} 0 & \text{if } J \neq S, \\ \operatorname{Det}(C) & \text{if } J = S. \end{cases} \tag{5.8.13}$$

The next exercise provides a relation that is similar to (5.8.13). To solve the exercise we have merely to repeat the above argument, but this time using Laplace's expansion in terms of columns rather than in terms of rows.

Exercise 4. *If Γ_{KS} is defined as in (5.8.12) show that*

$$\sum_K \Gamma_{JK} C_{KS} = \begin{cases} 0 & \textit{if } J \neq S, \\ \operatorname{Det}(C) & \textit{if } J = S. \end{cases} \tag{5.8.14}$$

In the preceding discussion J, K and S denoted increasing sequences of p integers all lying between 1 and n. The total number of such sequences is $\binom{n}{p}$. Let us put these $\binom{n}{p}$ sequences in some order – the particular way in which this is done does not matter. Then the ring elements C_{JK} can be regarded as the entries in an $\binom{n}{p}\times\binom{n}{p}$ matrix $C^{(p)}$ say. (This is the p-th exterior power of C.) Likewise the Γ_{JK} can be regarded as the entries in another $\binom{n}{p}\times\binom{n}{p}$ matrix $\Gamma^{(p)}$ say. On this understanding (5.8.13) and (5.8.14) can be written in matrix notation as

$$C^{(p)}\Gamma^{(p)}=(\mathrm{Det}(C))I=\Gamma^{(p)}C^{(p)}, \tag{5.8.15}$$

where I denotes the identity matrix with $\binom{n}{p}$ rows and columns.

Exercise 5*. *Let $C=\|c_{ij}\|$ be an $n\times n$ matrix with entries in R and for $1<p<n$ let $C^{(p)}$ be the p-th exterior power of C. Show that*

$$\mathrm{Det}(C^{(p)})=(\mathrm{Det}(C))^{(n-1,p-1)},$$

where by (r,s) is meant the binomial coefficient $r!/s!\,(r-s)!$.

To be fair to the reader it should be remarked that the solution provided for this exercise uses the fact that polynomials with integer coefficients form a unique factorization domain and also the result that the general determinant, considered as a polynomial in its entries, is irreducible.

We shall now interpret the result contained in Exercise 5 in terms of exterior algebras. To this end let F be a free R-module of rank n, and let $f: F\to F$ be an endomorphism of F. Further, suppose that $1<p<n$. If now $e_1, e_2, \ldots, e_n$ is a base for F, $f(e_r)=e_1c_{1r}+e_2c_{2r}+\cdots+e_nc_{nr}$, where $c_{ij}\in R$, and $C=\|c_{ij}\|$, then, of course, $\mathrm{Det}(f)=\mathrm{Det}(C)$. Let $J=(j_1, j_2, \ldots, j_p)$, where $1\le j_1<\cdots<j_p\le n$, and put $e_J=e_{j_1}\wedge e_{j_2}\wedge\cdots\wedge e_{j_p}$. Then the e_J form a base for $E_p(F)$ and, as we remarked earlier,

$$(E_p(f))(e_K)=\sum_J e_JC_{JK}.$$

But this means that, with respect to the base $\{e_J\}$ of $E_p(F)$, the endomorphism $E_p(f)$ has matrix $C^{(p)}$. Consequently $\mathrm{Det}(E_p(f))=\mathrm{Det}(C^{(p)})$. Exercise 5 now shows that

$$\mathrm{Det}(E_p(f))=(\mathrm{Det}(f))^{(n-1,p-1)} \tag{5.8.16}$$

which is the relation we were seeking.

We end this section with some comments on Pfaffians. The simplest alternating matrix of even order has the form

$$\begin{bmatrix} 0 & \alpha \\ -\alpha & 0 \end{bmatrix}$$

and the Pfaffian of this is α. At the end of Section (5.7) we calculated the Pfaffian of the general 4×4 alternating matrix. The reader may like to verify by the same method that the Pfaffian of

$$\begin{bmatrix} 0 & a_{12} & a_{13} & a_{14} & a_{15} & a_{16} \\ -a_{12} & 0 & a_{23} & a_{24} & a_{25} & a_{26} \\ -a_{13} & -a_{23} & 0 & a_{34} & a_{35} & a_{36} \\ -a_{14} & -a_{24} & -a_{34} & 0 & a_{45} & a_{46} \\ -a_{15} & -a_{25} & -a_{35} & -a_{45} & 0 & a_{56} \\ -a_{16} & -a_{26} & -a_{36} & -a_{46} & -a_{56} & 0 \end{bmatrix}$$

is

$$\begin{aligned} &a_{12}(a_{34}a_{56}-a_{35}a_{46}+a_{36}a_{45}) \\ &-a_{13}(a_{24}a_{56}-a_{25}a_{46}+a_{26}a_{45}) \\ &+a_{14}(a_{23}a_{56}-a_{25}a_{36}+a_{26}a_{35}) \\ &-a_{15}(a_{23}a_{46}-a_{24}a_{36}+a_{26}a_{34}) \\ &+a_{16}(a_{23}a_{45}-a_{24}a_{35}+a_{25}a_{34}). \end{aligned} \tag{5.8.17}$$

The general formula, of which this is a special case, will be derived later.

It will be noticed that the Pfaffian of an alternating matrix is a polynomial in its entries with *integer* coefficients. Let us make this more precise. Consider the *generic* alternating matrix

$$\begin{bmatrix} 0 & x_{12} & x_{13} & \cdots & x_{1n} \\ -x_{12} & 0 & x_{23} & \cdots & x_{2n} \\ -x_{13} & -x_{23} & 0 & \cdots & x_{3n} \\ \vdots & \vdots & \vdots & \vdots & \vdots \\ -x_{1n} & -x_{2n} & -x_{3n} & \cdots & 0 \end{bmatrix}$$

where the x_{ij} $(i<j)$ are $n(n-1)/2$ different indeterminants. We regard this matrix as having its entries in the polynomial ring $\mathbb{Z}[x_{12}, x_{13}, \ldots, x_{n-1,n}]$ and then its Pfaffian is a polynomial, say $Q(x_{12}, x_{13}, \ldots, x_{n-1,n})$, with integer coefficients in the indeterminates in question. Now let R be an arbitrary commutative ring and $A=\|a_{ij}\|$ an $n\times n$ alternating matrix with entries in R. Then

$$\mathrm{Pf}(A)=Q(a_{12}, a_{13}, \ldots, a_{n-1,n})$$

with the *same* polynomial Q as before. This follows from the method given in Section (5.7) for calculating Pfaffians.

For the next exercise we observe that if A and B are alternating matrices, then

$$\begin{bmatrix} A & 0 \\ 0 & B \end{bmatrix}$$

is an alternating matrix as well.

Exercise 6*. *Let A and B be alternating matrices with entries in R. Show that the Pfaffian of*

$$\begin{bmatrix} A & 0 \\ 0 & B \end{bmatrix}$$

is the product of the Pfaffian of A and the Pfaffian of B.

Now suppose that A is an $n\times n$ alternating matrix and that Λ is an arbitrary $n\times n$ matrix, both matrices having entries in R. If now Λ^T denotes the transpose of Λ, then it is easy to check that $\Lambda^T A\Lambda$ is also an alternating matrix, so that we can compare the Pfaffian of $\Lambda^T A \Lambda$ with that of A. The next exercise records the facts.

Exercise 7*. *Let A be an $n \times n$ alternating matrix and Λ an arbitrary $n \times n$ matrix. Show that $\Lambda^T A \Lambda$ is an alternating matrix and that*

$$\mathrm{Pf}(\Lambda^T A \Lambda) = (\mathrm{Det}(\Lambda))\,\mathrm{Pf}(A).$$

Let A be an $n \times n$ $(n > 2)$ alternating matrix and suppose that $1 \leq i < j \leq n$. If we strike out the i-th and j-th rows and also the i-th and j-th columns, then the resulting matrix will also be alternating.

Exercise 8*. *Let $A = \|a_{pq}\|$ be an $n \times n$ $(n > 2)$ alternating matrix and, for $1 \leq i < j \leq n$, let $\mathfrak{A}_{ij}$ be the matrix obtained from A by striking out the i-th and j-th rows and columns. Show that*

$$\mathrm{Pf}(A) = \sum_{j=2}^{n} (-1)^j a_{1j}\,\mathrm{Pf}(\mathfrak{A}_{1j}).$$

This result gives a convenient way of evaluating Pfaffians. We encountered a special case of it in (5.8.17).

Exercise 9. *Let I_n be the identity matrix of order n. Show that the Pfaffian of*

$$\begin{bmatrix} 0 & I_n \\ -I_n & 0 \end{bmatrix}$$

is equal to $(-1)^{n(n-1)/2}$

5.9 Solutions to selected exercises

Exercise 3. *Let $A = \|a_{ij}\|$ be an $n \times 2n$ matrix with entries in R, let $I = (i_1, i_2, \ldots, i_n)$ be a typical sequence of n integers satisfying $1 \leq i_1 < \cdots < i_n \leq 2n$, and let I' be the sequence obtained from $(1, 2, \ldots, 2n)$ by deleting the terms of I. If*

$$D_I = \begin{vmatrix} a_{1i_1} & a_{1i_2} & \cdots & a_{1i_n} \\ a_{2i_1} & a_{2i_2} & \cdots & a_{2i_n} \\ \vdots & \vdots & \vdots\vdots\vdots & \vdots \\ a_{ni_1} & a_{ni_2} & \cdots & a_{ni_n} \end{vmatrix}$$

show that

$$\sum_I \mathrm{sgn}(I, I') D_I D_{I'} = 0.$$

Show also that if k (where $1 \leq k \leq 2n$) is fixed and n is even, then

$$\sum_{k \in I} \mathrm{sgn}(I, I') D_I D_{I'} = 0.$$

Solution. Let C be the $2n \times 2n$ matrix

$$\begin{bmatrix} A \\ A \end{bmatrix}.$$

Then $\mathrm{Det}(C)=0$ because there are repeated rows. Hence if we use Laplace's expansion to evaluate the determinant (applying it to the first n rows) we find that

$$\sum_I (-1)^{1+2+\cdots+n}(-1)^{|I|} C_{(1,2,\ldots,n)I} C_{(n+1,n+2,\ldots,2n)I'} = 0$$

or

$$\sum_I \mathrm{sgn}(I, I') D_I D_{I'} = 0.$$

This is the first assertion. (Here we have used the fact that

$$\mathrm{sgn}(I, I') = (-1)^{(i_1-1)+\cdots+(i_n-n)}$$

which was noted previously.) Observe that

$$\mathrm{sgn}(I, I')\,\mathrm{sgn}(I', I) = (-1)^{|I|+|I'|} = (-1)^{n(2n+1)}.$$

Consequently

$$\mathrm{sgn}(I, I') = \mathrm{sgn}(I', I)$$

when n is even.

From here on we suppose that k $(1 \le k \le 2n)$ is a given integer which we keep fixed, and we suppose also that n is even. Let $I_1, I_2, \ldots, I_s$ be the sequences I that contain k. Then those that remain are $I'_1, I'_2, \ldots, I'_s$ and we have

$$\begin{aligned} 0 &= \sum \mathrm{sgn}(I, I') D_I D_{I'} \\ &= \sum_{t=1}^{s} \mathrm{sgn}(I_t, I'_t) D_{I_t} D_{I'_t} + \sum_{t=1}^{s} \mathrm{sgn}(I'_t, I''_t) D_{I'_t} D_{I''_t} \\ &= 2 \sum_{t=1}^{s} \mathrm{sgn}(I_t, I'_t) D_{I_t} D_{I'_t} \end{aligned}$$

because $I''_t = I_t$ and $\mathrm{sgn}(I_t, I'_t) = \mathrm{sgn}(I'_t, I_t)$. Thus

$$2 \sum_{k \in I} \mathrm{sgn}(I, I') D_I D_{I'} = 0 \tag{5.9.1}$$

and now we have the problem of getting rid of the 2.

We proceed as follows. Let

$$X = \begin{bmatrix} x_{11} & x_{12} & \cdots & x_{1,2n} \\ x_{21} & x_{22} & \cdots & x_{2,2n} \\ \vdots & \vdots & \vdots & \vdots \\ x_{n1} & x_{n2} & \cdots & x_{n,2n} \end{bmatrix}$$

where the x_{ij} are indeterminates, and consider X as a matrix with entries in $\mathbb{Z}[x_{11}, x_{12}, \ldots, x_{n,2n}]$. If now Δ_I denotes the subdeterminant of order n that comes from columns $i_1, i_2, \ldots, i_n$, then (5.9.1) shows that

$$2 \sum_{k \in I} \operatorname{sgn}(I, I')\Delta_I \Delta_{I'} = 0;$$

but now, because 2 is not a zerodivisor in the polynomial ring, we may conclude that

$$\sum_{k \in I} \operatorname{sgn}(I, I')\Delta_I \Delta_{I'} = 0. \tag{5.9.2}$$

Finally we observe that there is a ring-homomorphism of $\mathbb{Z}[x_{11}, x_{12}, \ldots, x_{n,2n}]$ into R in which m (in $\mathbb{Z}$) is mapped into $m1_R$ and x_{ij} is mapped into a_{ij}. If we apply this homomorphism to (5.9.2) we obtain the required relation. A convenient way to describe the last step is to say that

$$\sum_{k \in I} \operatorname{sgn}(I, I')D_I D_{I'} = 0$$

is obtained from (5.9.2) by *specialization.*

Exercise 5. *Let $C = \|c_{ij}\|$ be an $n \times n$ matrix with entries in R and for $1 < p < n$ let $C^{(p)}$ be the p-th exterior power of C. Show that*

$$\operatorname{Det}(C^{(p)}) = (\operatorname{Det}(C))^{(n-1, p-1)},$$

where by (r, s) is meant the binomial coefficient $r!/s!\,(s-r)!$.

Solution. Let

$$X = \begin{bmatrix} x_{11} & x_{12} & \cdots & x_{1n} \\ x_{21} & x_{22} & \cdots & x_{2n} \\ \vdots & \vdots & \vdots\vdots\vdots & \vdots \\ x_{n1} & x_{n2} & \cdots & x_{nn} \end{bmatrix}$$

where the x_{ij} are indeterminates and X is regarded as a matrix with entries in $\mathbb{Z}[x_{11}, x_{12}, \ldots, x_{nn}]$. We shall show that

$$\operatorname{Det}(X^{(p)}) = (\operatorname{Det}(X))^{(n-1, p-1)}.$$

Once this has been proved the desired result will follow by applying the ring-homomorphism

$$\mathbb{Z}[x_{11}, x_{12}, \ldots, x_{nn}] \to R$$

in which x_{ij} is mapped into c_{ij}.

By (5.8.15), we can find a square matrix U (say), of the same size as $X^{(p)}$, such that (i) the entries in U belong to $\mathbb{Z}[x_{11}, x_{12}, \ldots, x_{nn}]$, and (ii) $X^{(p)}U$ is $\operatorname{Det}(X)$ times an identity matrix. It follows that

$$\operatorname{Det}(X^{(p)})(\operatorname{Det}(U)) = (\operatorname{Det}(X))^{(n, p)}$$

and therefore $\mathrm{Det}(X^{(p)})$ is a factor, in the polynomial ring, of $(\mathrm{Det}(X))^{(n,p)}$. But $\mathrm{Det}(X)$ is irreducible in $\mathbb{Z}[x_{11}, x_{12}, \ldots, x_{nn}]$ and ± 1 are the only units in this ring. We now see that

$$\mathrm{Det}(X^{(p)}) = \varepsilon(\mathrm{Det}(X))^k,$$

where k is an integer and ε is either $+1$ or -1. By comparing degrees we find at once that

$$k = \frac{(n-1)!}{(p-1)!\,(n-p)!}.$$

To determine ε we give $x_{11}, x_{22}, \ldots, x_{nn}$ the value 1 and all the other x_{ij} the value zero. Then X specializes to the identity matrix I_n and we obtain $\varepsilon = \mathrm{Det}(I_n^{(p)})$. But $I_n^{(p)}$ is itself an identity matrix so that $\varepsilon = +1$.

Exercise 6. *Let A and B be alternating matrices with entries in R. Show that the Pfaffian of*

$$\begin{bmatrix} A & 0 \\ 0 & B \end{bmatrix} \tag{5.9.3}$$

is the product of the Pfaffian of A and the Pfaffian of B.

Solution. Let A have p rows and p columns and let B have q rows and q columns. Put $n = p + q$. Next, let M be a free R-module with a base $e_1, e_2, \ldots, e_n$ and let $\gamma: M \times M \to R$ be the alternating bilinear form in which $\|\gamma(e_i, e_j)\|$ is the matrix (5.9.3). Then, with the notation of Section (5.7), the desired Pfaffian is the component of degree zero belonging to

$$(L_{e_1} + \Delta_{e_1}) \circ (L_{e_2} + \Delta_{e_2}) \circ \cdots \circ (L_{e_n} + \Delta_{e_n})(1_{E(M)}).$$

Now

$$(L_{e_{p+1}} + \Delta_{e_{p+1}}) \circ (L_{e_{p+2}} + \Delta_{e_{p+2}}) \circ \cdots \circ (L_{e_n} + \Delta_{e_n})(1_{E(M)})$$
$$= \mathrm{Pf}(B) + e_{p+1} \wedge x_1 + e_{p+2} \wedge x_2 + \cdots + e_{p+q} \wedge x_q$$

for suitable elements $x_1, x_2, \ldots, x_q$ in $E(M)$. It is therefore enough to show that

$$(L_{e_1} + \Delta_{e_1}) \circ (L_{e_2} + \Delta_{e_2}) \circ \cdots \circ (L_{e_p} + \Delta_{e_p})\{e_{p+1} \wedge x_1 + \cdots + e_{p+q} \wedge x_q\}$$

has 0 as its component of degree zero. But this is clear because if $1 \le i \le p$, then $\Delta_{e_i}(e_{p+j}) = 0$ for $j = 1, 2, \ldots, q$, whereas operating with L_{e_i} replaces an element of the form $e_{p+1} \wedge x_1 + \cdots + e_{p+q} \wedge x_q$ by another element of the same form.

Exercise 7. *Let A be an $n \times n$ alternating matrix and let Λ be an arbitrary $n \times n$ matrix. Show that $\Lambda^T A \Lambda$ is an alternating matrix and that*

$$\mathrm{Pf}(\Lambda^T A \Lambda) = (\mathrm{Det}(\Lambda))\,\mathrm{Pf}(A).$$

Solution. Let X be the alternating matrix

$$\begin{bmatrix} 0 & x_{12} & x_{13} & \cdots & x_{1n} \\ -x_{12} & 0 & x_{23} & \cdots & x_{2n} \\ -x_{13} & -x_{23} & 0 & \cdots & x_{3n} \\ \vdots & \vdots & \vdots & \vdots\vdots\vdots & \vdots \\ -x_{1n} & -x_{2n} & -x_{3n} & \cdots & 0 \end{bmatrix}$$

and let

$$Y = \begin{bmatrix} y_{11} & y_{12} & \cdots & y_{1n} \\ y_{21} & y_{22} & \cdots & y_{2n} \\ \vdots & \vdots & \vdots\vdots\vdots & \vdots \\ y_{n1} & y_{n2} & \cdots & y_{nn} \end{bmatrix}.$$

Here the x_{ij} $(i<j)$ and the $y_{\mu\nu}$ are distinct indeterminates and X and Y are regarded as matrices with entries in the ring formed by adjoining all these indeterminates to the ring of integers. Now

$$(Y^T XY)^T = Y^T X^T Y = -(Y^T XY)$$

and so (because we are in characteristic zero) $Y^T XY$ is an alternating matrix. Note that it will suffice to show that $\mathrm{Pf}(Y^T XY) = (\mathrm{Det}(Y))\,\mathrm{Pf}(X)$ because the general result will then follow by specialization. But

$$\begin{aligned}(\mathrm{Pf}(Y^T XY))^2 &= \mathrm{Det}(Y^T XY)\\ &= (\mathrm{Det}(Y))^2\,\mathrm{Det}(X)\\ &= ((\mathrm{Det}(Y))\,\mathrm{Pf}(X))^2\end{aligned}$$

so either $\mathrm{Pf}(Y^T XY) = (\mathrm{Det}(Y))\,\mathrm{Pf}(X)$ or else $\mathrm{Pf}(Y^T XY) =$ $-(\mathrm{Det}(Y))\,\mathrm{Pf}(X)$. However, we have only to specialize Y so that it becomes the identity matrix to see that it is the former of these which is correct.

Exercise 8. *Let $A = \|a_{pq}\|$ be an $n \times n$ $(n>2)$ alternating matrix and, for $1 \le i < j \le n$, let $\mathfrak{A}_{ij}$ be the matrix obtained from A by striking out the i-th and j-th rows and columns. Show that*

$$\mathrm{Pf}(A) = \sum_{j=2}^{n} (-1)^j a_{1j}\,\mathrm{Pf}(\mathfrak{A}_{1j}).$$

Solution. It is convenient to deal with the problem in its generic form, that is we consider the matrix

$$X = \begin{bmatrix} 0 & x_{12} & x_{13} & \cdots & x_{1n} \\ -x_{12} & 0 & x_{23} & \cdots & x_{2n} \\ -x_{13} & -x_{23} & 0 & \cdots & x_{3n} \\ \vdots & \vdots & \vdots & \vdots\vdots\vdots & \vdots \\ -x_{1n} & -x_{2n} & -x_{3n} & \cdots & 0 \end{bmatrix}$$

where the x_{ij} $(i<j)$ are indeterminates and X is regarded as having its entries in $\mathbb{Z}[x_{12}, x_{13}, \ldots, x_{n-1,n}]$. Let Ξ_{ij} be the matrix obtained by striking out the i-th and j-th rows and columns. Then it will suffice to prove that

$$\mathrm{Pf}(X) = \sum_{j=2}^{n} (-1)^j x_{1j} \,\mathrm{Pf}(\Xi_{1j})$$

since all other cases will follow by specialization. Note that we can suppose that n is even for otherwise $\mathrm{Pf}(X)$ and the $\mathrm{Pf}(\Xi_{1j})$ will all be zero thus making the assertion trivial.

Suppose that $1 \le r \le n$. The indeterminates $x_{\lambda\mu}$ $(\lambda<\mu)$ with $r \in (\lambda, \mu)$ occur only in the r-th row and the r-th column. Each term of $\mathrm{Det}(X)$ is of degree 2 in these particular indeterminates, and therefore, since $\mathrm{Det}(X) = (\mathrm{Pf}(X))^2$, each term of $\mathrm{Pf}(X)$ will contain exactly one $x_{\lambda\mu}$ with $\lambda<\mu$ and $r \in (\lambda, \mu)$. It follows that

$$\mathrm{Pf}(X) = x_{12}U_2 + x_{13}U_3 + \cdots + x_{1n}U_n,$$

where U_j is a polynomial not involving any indeterminate x_{pq} $(p<q)$ with either 1 or j in (p, q).

Suppose that $2 \le j \le n$. Replace x_{pq} $(p<q)$ by zero if exactly one of 1 and j is in (p, q) and replace x_{1j} by 1. (All the other indeterminates are to remain unaltered.) Then X becomes a new alternating matrix, X_j say, and U_j is unchanged. We therefore have $U_j = \mathrm{Pf}(X_j)$.

Let Λ be the matrix obtained from the $n \times n$ identity matrix by taking its j-th column and moving it step by step to the left until it becomes the second column. Then $\mathrm{Det}(\Lambda) = (-1)^j$ and

$$\Lambda^T X_j \Lambda = \left[\begin{array}{cc|ccc} 0 & 1 & 0 & \cdots & 0 \\ -1 & 0 & 0 & \cdots & 0 \\ \hline \vdots & \vdots & & & \\ 0 & 0 & & \Xi_{1j} & \end{array}\right].$$

Consequently (see Exercises 6 and 7)

$$\mathrm{Pf}(\Xi_{1j}) = \mathrm{Pf}(\Lambda^T X_j \Lambda) = (\mathrm{Det}(\Lambda))\,\mathrm{Pf}(X_j)$$

whence $U_j = (-1)^j\,\mathrm{Pf}(\Xi_{1j})$. Accordingly

$$\mathrm{Pf}(X) = \sum_{j=2}^{n} (-1)^j x_{1j} \,\mathrm{Pf}(\Xi_{1j})$$

and the solution is complete.

6

The symmetric algebra of a module

General remarks

We have examined the origins and properties of the tensor and exterior algebras of a module. The *symmetric algebra*, which concerns us now, fits into the same general pattern and, as in the other cases, we shall motivate its investigation by means of an appropriate universal problem. However, from a different standpoint, one can say that the exterior algebra can be obtained from the tensor algebra by making it *anticommutative*. Viewed from this position, the symmetric algebra is the result of making the tensor algebra *commutative*.

Because much of the theory of the symmetric algebra runs closely parallel to that of the exterior algebra, we can frequently borrow proofs and adapt them without difficulty, and when this is the case we shall restrict the amount of detail that is given.

There is, however, one area in this account of symmetric algebras which is not foreshadowed in Chapter 5. The symmetric algebra of a module can be regarded as a generalization of a polynomial ring; and for a polynomial ring the partial differentiations with respect to the indeterminates generate a commutative algebra. When this observation is analysed it leads naturally to the algebra of *differential operators* and this is one of the topics we shall discuss. In the next chapter this particular algebra will be examined from a more general standpoint and, as a result, we shall be able to complete the analogy between exterior and symmetric algebras.

The general conventions that apply to this chapter are the same as those set out in the General Remarks at the commencement of Chapter 5 with one variation. Where we have to consider another commutative ring besides R we shall denote the second ring by R' rather than by S. This is because the letter S is widely used as part of the standard notation for symmetric algebras and symmetric powers.

6.1 The symmetric algebra

Let M be an R-module, A an R-algebra, and suppose that we are given an R-linear mapping $\phi: M \to A$ which is such that

$$\phi(m)\phi(m') = \phi(m')\phi(m) \tag{6.1.1}$$

for all m and m' in M. A natural universal problem connected with this situation can be posed as follows.

Problem. *To choose A and ϕ so that given any R-linear mapping $\psi: M \to B$ (B is an R-algebra) such that $\psi(m)\psi(m') = \psi(m')\psi(m)$ for all m, m' in M, there shall exist a unique homomorphism $h: A \to B$, of R-algebras, such that $h \circ \phi = \psi$.*

This type of problem will, by now, have become familiar. Clearly the problem has at most one solution in the sense that if (A, ϕ) and (A', ϕ') both meet the requirements, then there exist inverse isomorphisms $\lambda: A \to A'$ and $\lambda': A' \to A$, of R-algebras, such that $\lambda \circ \phi = \phi'$ and $\lambda' \circ \phi' = \phi$. As for the rest the essential facts are contained in the following theorem.

Theorem 1. *The universal problem just enunciated possesses a solution. Furthermore if (A, ϕ) is any solution, then*

(i) *$\phi: M \to A$ is an injection;*
(ii) *$\phi(M)$ generates A as an R-algebra;*
(iii) *there is a, necessarily unique, algebra-grading $\{A_n\}_{n\in\mathbb{Z}}$ on A with $A_1 = \phi(M)$;*
(iv) *for each $p \geq 1$, A_p is the p-th symmetric power of M, with $m_1 m_2 \ldots m_p = \phi(m_1)\phi(m_2)\ldots\phi(m_p)$ for elements $m_1, m_2, \ldots, m_p$ in M;*
(v) *the structural homomorphism $R \to A$ maps R isomorphically onto A_0.*

Remark. This result should be compared with Theorem 1 of Chapter 5. As will be seen the proofs are very similar.

Proof. Because any two solutions of the universal problem are copies of each other, it is sufficient to find *one* solution that satisfies all of (i)–(v).

Suppose that $m, m' \in M$. Then $m \otimes m' - m' \otimes m$ is a homogeneous element of degree two in the tensor algebra $T(M)$, and therefore such elements generate a homogeneous two-sided ideal, $H(M)$ say, in $T(M)$. Put $H_p(M) = H(M) \cap T_p(M)$ and $A = T(M)/H(M)$. Of course A has a natural structure as a graded R-algebra; let $\{A_n\}_{n\in\mathbb{Z}}$ be the grading which it inherits from $T(M)$. Then, in the natural homomorphism $T(M) \to A$ of graded algebras, $T_n(M)$ is mapped onto A_n. In particular, since $M = T_1(M)$, the homomorphism $T(M) \to A$ induces an R-linear mapping $\phi: M \to A$. Note that, because $m \otimes m' - m' \otimes m$ belongs to $H(M)$, its image in A is zero and therefore $\phi(m)\phi(m') = \phi(m')\phi(m)$.

It is a simple matter to check that A and ϕ satisfy conditions (i), (ii), (iii) and (v) in the statement of the theorem. To see that (iv) also holds suppose that $p \geq 1$. Then the surjective homomorphism $T_p(M) \to A_p$ induced by $T(M) \to A$ has kernel $T_p(M) \cap H(M) = H_p(M)$. Furthermore $H_p(M)$ is the submodule of $T_p(M)$ generated by all elements of the form

$$m_1 \otimes \cdots \otimes m_i \otimes m_{i+1} \otimes \cdots \otimes m_p$$
$$- m_1 \otimes \cdots \otimes m_{i+1} \otimes m_i \otimes \cdots \otimes m_p,$$

where the second term is the same as the first except that m_i and m_{i+1} have been interchanged. That A_p is the p-th symmetric power of M now follows from Theorem 8 of Chapter 1. Thus (iv) has been verified as well.

All that remains of the proof is for us to verify that (A, ϕ) solves the universal problem. Suppose then that $\psi: M \to B$ is an R-linear mapping, of M into an R-algebra B, and that $\psi(m)\psi(m') = \psi(m')\psi(m)$ for all m, m' in M. The extension of ψ to an algebra-homomorphism $T(M) \to B$ vanishes on $H(M)$ and so induces a homomorphism $h: A \to B$ of R-algebras. Clearly h satisfies $h \circ \phi = \psi$; and, since $\phi(M)$ generates A as an algebra, it is the only homomorphism of A into B to do so. This ends the proof.

We next introduce the standard notation and terminology. Let (A, ϕ) be a solution of our universal problem. We put

$$S(M) = A \tag{6.1.2}$$

and we employ $\{S_n(M)\}_{n \in \mathbb{Z}}$ to denote the grading referred to in Theorem 1. Also we use the fact that ϕ maps M isomorphically onto $A_1 = S_1(M)$ to make the identification

$$M = S_1(M), \tag{6.1.3}$$

and then M generates $S(M)$ as an R-algebra. Of course the grading on $S(M)$ is non-negative.

We shall use ordinary juxtaposition to denote multiplication in $S(M)$ so that, because $\phi(m)\phi(m') = \phi(m')\phi(m)$, we have from (6.1.3)

$$mm' = m'm \tag{6.1.4}$$

for m, m' in M. Next, when $p \geq 1$, $S_p(M)$ is the p-th symmetric power of M so naturally we call $S(M)$ the *symmetric algebra* of M. Finally Theorem 1 tells us that the structural homomorphism $R \to S(M)$ maps R isomorphically onto $S_0(M)$. It follows that we may make the identification $S_0(M) = R$ whenever it is convenient to do so.

Theorem 2. *Let $S(M)$ be the symmetric algebra of M and let $\psi: M \to B$ be an R-linear mapping of M into an R-algebra B. If now $\psi(m)\psi(m') = \psi(m')\psi(m)$*

for all m, m' in M, then ψ has a unique extension to a homomorphism $S(M) \to B$ of R-algebras.

Of course Theorem 2 holds because of the connection between $S(M)$ and the universal problem with which we started.

The proof of Theorem 1 reveals the relation of the symmetric algebra to the tensor algebra. However, we shall now describe this relation in a less encumbered context. To this end observe that the inclusion mapping $M \to S(M)$ extends to a surjective homomorphism

$$T(M) \to S(M) \tag{6.1.5}$$

of graded R-algebras, and so for each $p \in \mathbb{Z}$ there is induced a surjective homomorphism

$$T_p(M) \to S_p(M) \tag{6.1.6}$$

of R-modules. If $p \geq 1$, then the mapping (6.1.6) takes $m_1 \otimes m_2 \otimes \cdots \otimes m_p$ into $m_1 m_2 \ldots m_p$ and therefore it is none other than the canonical homomorphism which we first met in Section (1.5). Thus (6.1.5) is effectively a combination of canonical homomorphisms.

The next result is the counterpart of Theorem 3 of Chapter 5.

Theorem 3. *The symmetric algebra $S(M)$ is commutative.*

This follows from (6.1.4) and the fact that M generates $S(M)$ as an R-algebra.

6.2 Functorial properties

The functorial properties of the symmetric algebra now follow in a straightforward manner. Thus if $f: M \to N$ is a homomorphism of R-modules, then $N = S_1(N)$ by (6.1.3) and therefore (see Theorem 2) f can be extended to a homomorphism

$$S(f): S(M) \to S(N) \tag{6.2.1}$$

of R-algebras. But $S(f)$ preserves the degrees of homogeneous elements. Consequently for each integer p there is induced an R-linear mapping

$$S_p(f): S_p(M) \to S_p(N). \tag{6.2.2}$$

Note that $S_1(f) = f$ and that if we make the identifications $S_0(M) = R = S_0(N)$, then $S_0(f)$ is the identity mapping of R. Note too that if $p \geq 1$ and $m_1, m_2, \ldots, m_p$ belong to M, then

$$S_p(f)(m_1 m_2 \ldots m_p) = f(m_1) f(m_2) \ldots f(m_p). \tag{6.2.3}$$

Of course, if f is the identity mapping of M, then $S(f)$ is the identity mapping of $S(M)$. Also if $f: M \to N$ and $g: N \to K$ are homomorphisms of R-modules, then

$$S(g \circ f) = S(g) \circ S(f) \tag{6.2.4}$$

and hence, for all $p \in \mathbb{Z}$,

$$S_p(g \circ f) = S_p(g) \circ S_p(f). \tag{6.2.5}$$

Thus $S(M)$ can be regarded as a covariant functor from R-modules to commutative graded R-algebras. Naturally if f is an isomorphism of R-modules, then $S(f)$ is an isomorphism of graded R-algebras and

$$S(f)^{-1} = S(f^{-1}). \tag{6.2.6}$$

6.3 The symmetric algebra of a free module

Throughout Section (6.3) we shall assume that the ring R is non-trivial.

Let M be a free R-module with a base B. If $p \geq 1$, then (Chapter 1, Theorem 9) $S_p(M)$ is a free module and it has a base consisting of the different products $b_1 b_2 \ldots b_p$, where $b_i \in B$. We also know that two such products, say $b_1 b_2 \ldots b_p$ and $b'_1 b'_2 \ldots b'_p$, are equal if and only if $(b'_1, b'_2, \ldots, b'_p)$ is a rearrangement of $(b_1, b_2, \ldots, b_p)$. Of course $S_0(M)$ is a free R-module having the identity element of $S(M)$ as a base.

The above remarks show that $S(M)$ itself, considered as an R-module, is free and has as a base the set of all distinct products $b_1 b_2 \ldots b_n$, where $b_i \in B$ and $n \geq 0$; moreover two products $b_1 b_2 \ldots b_n$ and $b'_1 b'_2 \ldots b'_q$ are the same if and only if $n = q$ and $(b'_1, b'_2, \ldots, b'_q)$ is a rearrangement of $(b_1, b_2, \ldots, b_n)$. Since the multiplicative structure of $S(M)$ is determined by the way the basis elements are to be multiplied, it follows that

$$S(M) = R[B]. \tag{6.3.1}$$

Here $R[B]$ denotes the ordinary ring of polynomials, with coefficients in R, in which the elements of B are treated as distinct commuting indeterminates.

Theorem 4. *Let P be a projective R-module. Then $S(P)$, considered as an R-module, is also projective. Hence, for each $n \in \mathbb{Z}$, $S_n(P)$ is a projective R-module.*

Proof. This result resembles Theorem 5 of Chapter 4, and there should be no difficulty in adapting the proof of the earlier theorem to meet the needs of the present one.

6.4 The symmetric algebra of a direct sum

If $A_1, A_2, \ldots, A_q$ are *commutative* R-algebras, then $A_1 \otimes A_2 \otimes \cdots \otimes A_q$ is also commutative. Hence if $M_1, M_2, \ldots, M_q$ are R-modules, then

$$S(M_1) \otimes S(M_2) \otimes \cdots \otimes S(M_q) \tag{6.4.1}$$

is a graded, commutative R-algebra. (Note that here we are concerned with the ordinary and not the modified tensor product.) The elements of degree one in this algebra form the module

$$M_1 \otimes R \otimes \cdots \otimes R + R \otimes M_2 \otimes \cdots \otimes R + \cdots + R \otimes R \otimes \cdots \otimes M_q, \tag{6.4.2}$$

where each term in the sum contains q factors; moreover this module generates $S(M_1) \otimes S(M_2) \otimes \cdots \otimes S(M_q)$ as an R-algebra.

Consider the R-linear mapping

$$\phi: M_1 \oplus M_2 \oplus \cdots \oplus M_q \to S(M_1) \otimes S(M_2) \otimes \cdots \otimes S(M_q), \tag{6.4.3}$$

where

$$\phi(m_1, m_2, \ldots, m_p) = \sum_{i=1}^{q} 1 \otimes \cdots \otimes m_i \otimes \cdots \otimes 1 \tag{6.4.4}$$

it being understood that in $1 \otimes \cdots \otimes m_i \otimes \cdots \otimes 1$ the factor m_i occurs in the i-th position.

Theorem 5. *The tensor product*

$$S(M_1) \otimes_R S(M_2) \otimes_R \cdots \otimes_R S(M_q) \tag{6.4.5}$$

together with the R-linear mapping

$$\phi: M_1 \oplus M_2 \oplus \cdots \oplus M_q \to S(M_1) \otimes_R \cdots \otimes_R S(M_q)$$

constitute the symmetric algebra of $M_1 \oplus M_2 \oplus \cdots \oplus M_q$. *Furthermore the grading which* (6.4.5) *possesses by virtue of being a tensor product of graded algebras is the same as its grading when considered as the symmetric algebra of* $M_1 \oplus M_2 \oplus \cdots \oplus M_q$.

Proof. This result should be compared with Theorem 8 of Chapter 5 and once again it is a simple matter to adapt the proof of the corresponding result for exterior algebras. For instance it is possible to reduce the general result to the case where $q=2$. The only point which possibly deserves mention is that whereas in Chapter 5 we used the isomorphism

$$(E(M_1) \underline{\otimes} \cdots \underline{\otimes} E(M_{q-1})) \underline{\otimes} E(M_q) \approx E(M_1) \underline{\otimes} E(M_2) \underline{\otimes} \cdots \underline{\otimes} E(M_q)$$

this time all we need is the less sophisticated

$$(S(M_1) \otimes \cdots \otimes S(M_{q-1})) \otimes S(M_q) \approx S(M_1) \otimes S(M_2) \otimes \cdots \otimes S(M_q)$$

which, of course, is a special case of (3.3.13).

6.5 Covariant extension of a symmetric algebra

Let $\omega: R \to R'$ be a homomorphism of commutative rings and let M be an R-module. In this section the symmetric algebra of M will be denoted by $S_R(M)$. We note that $S_R(M) \otimes_R R'$ is a *commutative* R'-algebra and that it is graded by the family $\{S_n(M) \otimes_R R'\}_{n \in \mathbb{Z}}$ of R'-submodules. The homogeneous elements of degree one in $S_R(M) \otimes_R R'$ form the covariant extension $M \otimes_R R'$ of M, and $M \otimes_R R'$ generates $S_R(M) \otimes_R R'$ as an R'-algebra.

Let $S_{R'}(M \otimes_R R')$ denote the symmetric algebra of $M \otimes_R R'$ considered as an R'-module. Then, because $S_R(M) \otimes_R R'$ is commutative, the inclusion mapping $M \otimes_R R' \to S_R(M) \otimes_R R'$ extends to a homomorphism

$$\lambda: S_{R'}(M \otimes_R R') \to S_R(M) \otimes_R R' \tag{6.5.1}$$

of R'-algebras which evidently preserves degrees.

Theorem 6. *With the above notation*

$$\lambda: S_{R'}(M \otimes_R R') \to S_R(M) \otimes_R R'$$

is an isomorphism of graded R'-algebras.

We leave to the reader the straightforward task of adapting the proof of Theorem 6 of Chapter 4. Finally we observe that Theorem 6 allows us to make the identifications

$$S_{R'}(M \otimes_R R') = S_R(M) \otimes_R R' \tag{6.5.2}$$

and (for all $n \in \mathbb{Z}$)

$$S_n(M \otimes_R R') = S_n(M) \otimes_R R'. \tag{6.5.3}$$

6.6 Derivations on a symmetric algebra

Let M be an R-module and let us identify the submodule $S_0(M)$ of the symmetric algebra with R. If now D is a derivation on $S(M)$, then D induces a homomorphism of $S_1(M) = M$ into $S_0(M) = R$, so that in effect D extends a linear form on M.

Theorem 7. *Let f be a linear form on M. Then there is one and only one derivation on $S(M)$ that extends f.*

Proof. Lemma 4 of Chapter 3 shows that at most one derivation extends f. Now suppose that $p \geq 1$. There is a *symmetric* multilinear mapping, of the p-fold product $M \times M \times \cdots \times M$ into $S_{p-1}(M)$, in which $(m_1, m_2, \ldots, m_p)$ becomes

$$\sum_{i=1}^{p} f(m_i) m_1 \ldots \hat{m}_i \ldots m_p,$$

the ^ over m_i indicating, as usual, that this term is to be omitted. It follows that there exists an R-homomorphism $D_p: S_p(M) \to S_{p-1}(M)$ that satisfies

$$D_p(m_1 m_2 \ldots m_p) = \sum_{i=1}^{p} f(m_i) m_1 \ldots \hat{m}_i \ldots m_p.$$

Next, by combining $D_1, D_2, D_3, \ldots$ we obtain an R-endomorphism $D: S(M) \to S(M)$, of degree -1, that agrees with D_p on $S_p(M)$. A simple verification now shows that D is a derivation. It extends f because $D_1 = f$.

6.7 Differential operators

The notion of a *differential operator* is one that we have not encountered before. It occurs naturally in connection with polynomial rings and it can be extended readily to symmetric algebras. We shall therefore investigate it here.

Let M be an R-module and h a *non-negative* integer. A differential operator of degree h, on $S(M)$, is an R-linear mapping $\phi: S(M) \to S(M)$ of degree $-h$ with a certain additional property. To describe the property it will be useful to introduce some notation.

Suppose that $p \geq h$ and that $m_1, m_2, \ldots, m_p$ belong to M. Let $I = (i_1, i_2, \ldots, i_h)$ be a sequence of h integers with $1 \leq i_1 < i_2 < \cdots < i_h \leq p$, and let I' be the increasing sequence obtained from $(1, 2, \ldots, p)$ by deleting the terms in I. Finally put

$$m_I = m_{i_1} m_{i_2} \ldots m_{i_h} \tag{6.7.1}$$

and define $m_{I'}$ similarly.

On this understanding an R-endomorphism $\phi: S(M) \to S(M)$ of degree $-h$ is called a *differential operator of degree h* provided that

$$\phi(m_1 m_2 \ldots m_p) = \sum_I \phi(m_I) m_{I'} \tag{6.7.2}$$

for all $p \geq h$ and arbitrary elements $m_1, m_2, \ldots, m_p$ in M. It is clear that the differential operators of a given degree h form an R-submodule of $\mathrm{End}_R(S(M))$. This submodule will be denoted by $\nabla_h S(M)$. Of course, $\nabla_0 S(M)$ consists of the endomorphisms of $S(M)$ produced by multiplication by the various elements of R. An explanation of the name 'differential operator' will be given in Section (6.8). Note that a differential operator of degree 1 is identical with a *derivation* as defined in Section (3.7).

There is a second form of (6.7.2) to which it is convenient to draw attention. Suppose that $x_1, x_2, \ldots, x_q$ are elements of M and that $s_1, s_2, \ldots, s_q$ are non-negative integers with $s_1 + s_2 + \cdots + s_q \geq h$. If we replace $m_1 m_2 \ldots m_p$ by $x_1^{s_1} x_2^{s_2} \ldots x_q^{s_q}$, then (6.7.2) becomes

$$\phi(x_1^{s_1} x_2^{s_2} \ldots x_q^{s_q}) = \sum \binom{s_1}{\alpha_1} \ldots \binom{s_q}{\alpha_q} x_1^{s_1 - \alpha_1} x_2^{s_2 - \alpha_2} \ldots x_q^{s_q - \alpha_q}, \tag{6.7.3}$$

where the summation is over all sequences $(\alpha_1, \alpha_2, \ldots, \alpha_q)$ of integers with

$$0 \leq \alpha_i \leq s_i \quad (i = 1, 2, \ldots, q) \tag{6.7.4}$$

and

$$\alpha_1+\alpha_2+\cdots+\alpha_q=h. \tag{6.7.5}$$

Note that we have allowed the possibility that some of $s_1, s_2, \ldots, s_q$ may be zero and that this involves the use of appropriate conventions to cover this situation.

Lemma 1. *Let $h \geq 0$ and $k \geq 0$ be integers, and let ϕ and ψ be differential operators, on $S(M)$, of degrees h and k respectively. Then $\psi \circ \phi = \phi \circ \psi$ and this is a differential operator of degree $h+k$.*

Proof. Suppose that $p \geq h+k$ and that $m_1, m_2, \ldots, m_p$ belong to M. In what follows $I=(i_1, i_2, \ldots, i_h)$, $J=(j_1, j_2, \ldots, j_k)$ and $L=(l_1, l_2, \ldots, l_{h+k})$ denote strictly increasing sequences of integers, between 1 and p, whose lengths are h, k and $h+k$ respectively. We shall use (I, J) to denote the sequence obtained from $(i_1, i_2, \ldots, j_1, j_2, \ldots)$ by arranging the *different* integers present so that they form an increasing sequence. We defined I' above. L' is defined similarly.

It is clear that $\psi \circ \phi$ and $\phi \circ \psi$ are endomorphisms of degree $-(h+k)$, and from (6.7.2) we see that

$$(\psi \circ \phi)(m_1 m_2 \ldots m_p) = \sum_L \left\{ \sum_{(I,J)=L} \phi(m_I)\psi(m_J) \right\} m_{L'}. \tag{6.7.6}$$

This shows that $\psi \circ \phi = \phi \circ \psi$. Again

$$(\psi \circ \phi)(m_L) = \sum_{(I,J)=L} \phi(m_I)\psi(m_J)$$

and now we see that $\psi \circ \phi$ is a differential operator.

A differential operator of degree h ($h \geq 0$) induces a linear form on $S_h(M)$ and (6.7.2) shows that if we know h and the associated linear form on $S_h(M)$, then the differential operator is fully determined. The next theorem shows that the linear form may be prescribed arbitrarily.

Theorem 8. *Let h ($h \geq 0$) be a given integer, let M be an R-module, and let f be a linear form on $S_h(M)$. Then there is exactly one differential operator of degree h, on $S(M)$, which extends f.*

Proof. Suppose that $p \geq h$ and consider the mapping of the p-fold product $M \times M \times \cdots \times M$ into $S_{p-h}(M)$ in which $(m_1, m_2, \ldots, m_p)$ becomes

$$\sum_I f(m_I) m_{I'}.$$

(Here $I=(i_1, i_2, \ldots, i_h)$ satisfies $1 \leq i_1 < \cdots < i_h \leq p$ and m_I and $m_{I'}$ have the same meanings as in (6.7.2).) This mapping is both multilinear and symmetric and so it induces an R-linear mapping $\phi_p\colon S_p(M) \to S_{p-h}(M)$

with the property that

$$\phi_p(m_1 m_2 \dots m_p) = \sum_I f(m_I) m_{I'}.$$

Thus far we have assumed that $p \geq h$. For $0 \leq p < h$ we take ϕ_p to be the null homomorphism of $S_p(M)$ into $S_{p-h}(M)$.

We can now combine $\phi_0, \phi_1, \phi_2, \dots$ so as to obtain an endomorphism $\phi: S(M) \to S(M)$ of degree $-h$ which agrees with ϕ_q on $S_q(M)$. By construction, ϕ is a differential operator of degree h and it extends f. Finally, we already know that there can be no other differential operator with these properties.

Corollary. *Let h be a non-negative integer. Then the module $\nabla_h S(M)$ of differential operators of degree h is isomorphic to the module of linear forms on $S_h(M)$ under an isomorphism which matches ϕ, in $\nabla_h S(M)$, with its restriction to $S_h(M)$.*

Of course $\nabla_h S(M)$ is an R-submodule of $\mathrm{End}_R(S(M))$. Put

$$\nabla S(M) = \sum_{h \geq 0} \nabla_h S(M), \tag{6.7.7}$$

where the sum is taken in $\mathrm{End}_R(S(M))$. It is easy to see that this sum is *direct*. Indeed in view of Lemma 1 we have

Theorem 9. *$\nabla S(M)$ is a commutative R-subalgebra of* $\mathrm{End}_R(S(M))$ *and it is non-negatively graded by its submodules $\{\nabla_h S(M)\}_{h \geq 0}$.*

Definition. *The graded, commutative algebra $\nabla S(M)$ is called the 'algebra of differential operators' on $S(M)$.*

In the next chapter we shall meet this algebra in a different guise.

6.8 Comments and exercises

We make some miscellaneous comments on the subject of symmetric algebras. As always, if M is an R-module, then $S(M)$ – or $S_R(M)$ if we wish to be more explicit – denotes its symmetric algebra.

Let us consider what happens to $S(M)$ if we factor out an ideal of R. Suppose then that I is an R-ideal. Then $IS(M)$ is a homogeneous, two-sided ideal of $S(M)$ and hence $S(M)/IS(M)$ is a graded algebra both with respect to R and with respect to the ring R/I. The homomorphism of $M = S_1(M)$ that is induced by the natural mapping $S(M) \to S(M)/IS(M)$ vanishes on IM and so there results an R/I-linear mapping

$$M/IM \to S(M)/IS(M). \tag{6.8.1}$$

But $S(M)/IS(M)$ is a *commutative* R/I-algebra and therefore (6.8.1) extends to a homomorphism

$$\phi: S_{R/I}(M/IM) \to S(M)/IS(M) \tag{6.8.2}$$

of R/I-algebras.

The next exercise should be compared with Exercise 1 of Chapter 5.

Exercise 1. *Show that the homomorphism (6.8.2) is an isomorphism of graded R/I-algebras.*

This exercise shows that

$$S_{R/I}(M/IM) = S_R(M)/IS_R(M). \tag{6.8.3}$$

Exercise 2. *Let M be an R-module, let i be a given integer, and let $f: M \to S_{i+1}(M)$ be an R-linear mapping. Show that there is one and only one generalized derivation of degree i, on $S(M)$, that extends f.*

This, of course, is the counterpart of Exercise 2 of Chapter 5. It can be solved in very much the same way.

We turn now to the consideration of differential operators and we begin by indicating the origin of the name. First, however, it will be convenient to introduce some additional notation.

Let $\alpha = (\alpha_1, \alpha_2, \ldots, \alpha_n)$ and $\beta = (\beta_1, \beta_2, \ldots, \beta_n)$ be two sequences of n non-negative integers and let us put

$$|\alpha| = \alpha_1 + \alpha_2 + \cdots + \alpha_n \tag{6.8.4}$$

and

$$\binom{\beta}{\alpha} = \binom{\beta_1}{\alpha_1}\binom{\beta_2}{\alpha_2}\cdots\binom{\beta_n}{\alpha_n}. \tag{6.8.5}$$

We shall use $\alpha \leq \beta$ to mean that $\alpha_i \leq \beta_i$ for $i = 1, 2, \ldots, n$ so that, on this understanding, $\binom{\beta}{\alpha} = 0$ whenever $\alpha \nleq \beta$. Lastly, $\alpha + \beta$ and $\alpha - \beta$ will denote the sequences $(\alpha_1 + \beta_1, \ldots, \alpha_n + \beta_n)$ and $(\alpha_1 - \beta_1, \ldots, \alpha_n - \beta_n)$ respectively.

Now let M be a free R-module having $X_1, X_2, \ldots, X_n$ as a base. Then, by (6.3.1), $S(M)$ is the polynomial ring $R[X_1, X_2, \ldots, X_n]$ and, if $p \geq 0$, $S_p(M)$ consists of all the homogeneous polynomials, i.e. *forms*, of degree p. It will be convenient to denote the power-product $X_1^{\alpha_1} X_2^{\alpha_2} \ldots X_n^{\alpha_n}$ by X^α. We can then say that $S_p(M)$ is a free R-module having the X^α with $|\alpha| = p$ as a base.

Suppose next that $\phi: S(M) \to S(M)$ is an R-linear mapping of degree $-h$ $(h \geq 0)$. By (6.7.3) this will be a differential operator of degree h provided that, whenever $|\beta| \geq h$, we have

$$\phi(X^\beta) = \sum \binom{\beta}{\alpha} \phi(X^\alpha) X^{\beta - \alpha}, \tag{6.8.6}$$

where the sum is taken over all α with $\alpha \leq \beta$ and $|\alpha| = h$. It is now apparent that if $\gamma_i \geq 0$ for $i = 1, 2, \ldots, n$, then

$$\frac{\partial^{|\gamma|}}{\partial X_1^{\gamma_1}\,\partial X_2^{\gamma_2}\ldots\partial X_n^{\gamma_n}}$$

is a differential operator of degree $|\gamma|$. It is this fact which accounts for the terminology.

Suppose now that $p\geq 0$. We know that $S_p(M)$ has $\{X^\beta\}_{|\beta|=p}$ as a base. For $|\alpha|=p=|\beta|$, let $D^{(\alpha)}$ be the linear form on $S_p(M)$ which satisfies

$$D^{(\alpha)}(X^\beta)=\begin{cases}0 & \text{if } \alpha\neq\beta,\\ 1 & \text{if } \alpha=\beta.\end{cases}$$

Then the $D^{(\alpha)}$ form a base for the linear forms on $S_p(M)$. But, by Theorem 8, $D^{(\alpha)}$ has a unique extension to a differential operator of degree p on $S(M)$; let us use $D^{(\alpha)}$ also to denote the differential operator. We now see not only that $\nabla_p S(M)$ is a free R-module, but also that it has $\{D^{(\alpha)}\}_{|\alpha|=p}$ as a base. Note that for every sequence $\gamma=(\gamma_1,\gamma_2,\ldots,\gamma_n)$ of non-negative integers, we have

$$D^{(\alpha)}(X^\gamma)=\begin{cases}\binom{\gamma}{\alpha}X^{\gamma-\alpha} & \text{if } \alpha\leq\gamma,\\ 0 & \text{if } \alpha\nleq\gamma.\end{cases}$$

This formula makes it clear that

$$\frac{\partial^{|\alpha|}}{\partial X_1^{\alpha_1}\,\partial X_2^{\alpha_2}\ldots\partial X_n^{\alpha_n}}=(\alpha_1!)(\alpha_2!)\ldots(\alpha_n!)D^{(\alpha)}.$$

Exercise 3. *With the above notation show that*

$$D^{(\alpha)}\circ D^{(\beta)}=\binom{\alpha+\beta}{\alpha}D^{(\alpha+\beta)}=D^{(\beta)}\circ D^{(\alpha)}.$$

We turn now to other matters. Let U and M be R-modules and suppose that we are given a bilinear form

$$\gamma\colon U\times M\to R. \tag{6.8.7}$$

For a fixed u, in U, the mapping $m\mapsto\gamma(u,m)$ is a linear form on M and this, by Theorem 7, will have a unique extension, D_u say, to a derivation on $S(M)$. Thus, by construction,

$$D_u(m)=\gamma(u,m) \tag{6.8.8}$$

for all $m\in M$. Next, the mapping $U\to\mathrm{End}_R(S(M))$ given by $u\mapsto D_u$ is R-linear and, by Lemma 4 of Chapter 3, $D_{u_1}\circ D_{u_2}=D_{u_2}\circ D_{u_1}$ for all u_1, u_2 in U. It follows that there is a homomorphism

$$\Gamma\colon S(U)\to\mathrm{End}_R(S(M)) \tag{6.8.9}$$

of R-algebras which has the property that

$$\Gamma(u_1u_2\ldots u_p)=D_{u_1}\circ D_{u_2}\circ\cdots\circ D_{u_p} \tag{6.8.10}$$

for $u_1, u_2, \ldots, u_p$ in U. But, because a derivation is a differential operator of degree 1, $D_{u_1}\circ D_{u_2}\circ\cdots\circ D_{u_p}$ is a differential operator of degree p. Hence

(6.8.9) *is a degree-preserving, algebra-homomorphism of* $S(U)$ *into the algebra of differential operators on* $S(M)$.

Let us examine, in more detail, how this homomorphism operates. First we recall that if

$$C = \begin{bmatrix} c_{11} & c_{12} & \cdots & c_{1n} \\ c_{21} & c_{22} & \cdots & c_{2n} \\ \vdots & \vdots & \vdots\vdots & \vdots \\ c_{n1} & c_{n2} & \cdots & c_{nn} \end{bmatrix}$$

is an $n \times n$ matrix, then its *permanent*, $\mathrm{Per}(C)$, is defined by

$$\mathrm{Per}(C) = \sum_{\pi} c_{1\pi(1)} c_{2\pi(2)} \ldots c_{n\pi(n)},$$

where π is a typical permutation of $(1, 2, \ldots, n)$. The next exercise uses this concept to describe the effect, on $m_1 m_2 \ldots m_p$, of the differential operator associated with $u_1 u_2 \ldots u_p$.

Exercise 4. *Let* $u_1, u_2, \ldots, u_p$ *belong to* U *and* $m_1, m_2, \ldots, m_p$ *to* M. *Show that* $(D_{u_1} \circ D_{u_2} \circ \cdots \circ D_{u_p})(m_1 m_2 \ldots m_p)$ *is equal to the permanent of the matrix*

$$\begin{bmatrix} \gamma(u_1, m_1) & \gamma(u_1, m_2) & \cdots & \gamma(u_1, m_p) \\ \gamma(u_2, m_1) & \gamma(u_2, m_2) & \cdots & \gamma(u_2, m_p) \\ \vdots & \vdots & \vdots\vdots & \vdots \\ \gamma(u_p, m_1) & \gamma(u_p, m_2) & \cdots & \gamma(u_p, m_p) \end{bmatrix}.$$

7

Coalgebras and Hopf algebras

General remarks

A *coalgebra* is a concept which is dual (in a sense that belongs to Category Theory) to that of an associative algebra, and consequently to almost every result that we have concerning algebras there is a corresponding result for coalgebras. Now it sometimes happens that an algebra and a coalgebra are built on the same underlying set. When this occurs, and provided the algebra and the coalgebra interact suitably, the result is called a *Hopf* algebra. Our concern with these matters stems from the fact that, when M is an R-module, both $E(M)$ and $S(M)$ are Hopf algebras of particular interest.

Whenever we have a coalgebra the linear forms on it can be considered as the elements of an algebra. The algebra which arises in this way from $E(M)$ is known as the *Grassmann* algebra of M; for $S(M)$ the resulting algebra has very close connections with the algebra of differential operators which was described in Chapter 6.

Throughout this chapter we shall follow our usual practice of using R to denote a commutative ring with an identity element; and when the tensor symbol $\otimes$ is used it is understood that the underlying ring is always R unless there is an explicit statement indicating the contrary.

Finally, it happens to be convenient, before introducing the notion of a coalgebra, to reformulate the definition of an R-algebra. This reformulation is carried out in Section (7.1).

7.1 A fresh look at algebras

Let A be an associative R-algebra with an identity element. Then *inter alia* A is an R-module. Further, we have an R-linear mapping $\mu: A \otimes A \to A$ which is such that $\mu(x \otimes y) = xy$ and this, because multiplication on A is associative, makes

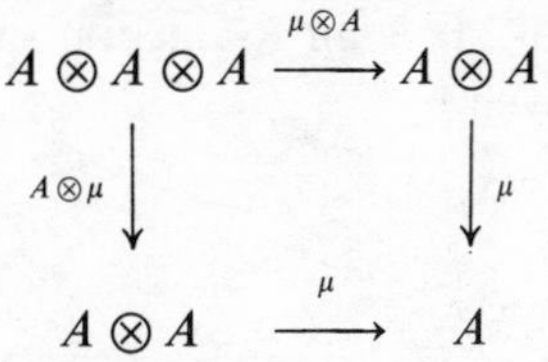

$$\begin{array}{ccc} A\otimes A\otimes A & \xrightarrow{\mu\otimes A} & A\otimes A \\ {\scriptstyle A\otimes\mu}\downarrow & & \downarrow{\scriptstyle \mu} \\ A\otimes A & \xrightarrow{\mu} & A \end{array} \tag{7.1.1}$$

a commutative diagram.

Now let $\eta: R \to A$ be the structural homomorphism of the algebra A. This too is R-linear and, because $\eta(1)$ is the identity element of A, the diagram

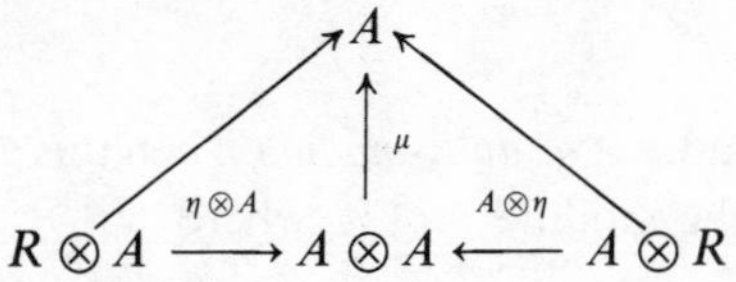

$$\begin{array}{ccccc} & & A & & \\ & \nearrow & \uparrow{\scriptstyle \mu} & \nwarrow & \\ R\otimes A & \xrightarrow{\eta\otimes A} & A\otimes A & \xleftarrow{A\otimes\eta} & A\otimes R \end{array} \tag{7.1.2}$$

also commutes. (Here $R \otimes A \to A$ and $A \otimes R \to A$ are the isomorphisms provided by Theorem 3 of Chapter 2.)

At this point we make a fresh start. Suppose that A is an R-*module* and that we are given R-linear mappings $\mu: A \otimes A \to A$ and $\eta: R \to A$ which are such that (7.1.1) and (7.1.2) are commutative diagrams. For x, y in A put $xy = \mu(x \otimes y)$. Then it is easily verified that this definition of multiplication turns A into an R-algebra that has $\eta(1)$ as its identity element and $\eta: R \to A$ as its structural homomorphism.

When algebras are looked at in this way we shall say that the triple (A, μ, η) constitutes an associative R-algebra with identity, and (in this context) μ is called the *multiplication mapping* and η the *unit mapping* of the algebra. Let us now review the general aspects of the theory of algebras from this new standpoint.

Suppose that (A, μ_A, η_A) and (B, μ_B, η_B) are (associative) R-algebras and let f be a mapping of A into B. It is readily checked that f is an algebra-homomorphism, in the sense of Section (3.1), if and only if the conditions (i) f is R-linear, (ii) $f \circ \mu_A = (f \otimes f) \circ \mu_B$, and (iii) $f \circ \eta_A = \eta_B$ are all satisfied.

Now assume that, for $1 \leq i \leq p$, (A_i, μ_i, η_i) is an R-algebra. Let

$$\begin{aligned} \Lambda_p: (A_1 \otimes \cdots \otimes A_p) &\otimes (A_1 \otimes \cdots \otimes A_p) \\ &\xrightarrow{\sim} (A_1 \otimes A_1) \otimes \cdots \otimes (A_p \otimes A_p) \end{aligned} \tag{7.1.3}$$

be the isomorphism (of R-modules) which matches $(a_1 \otimes \cdots \otimes a_p) \otimes (a_1' \otimes \cdots \otimes a_p')$ with $(a_1 \otimes a_1') \otimes \cdots \otimes (a_p \otimes a_p')$. Further, let

$$\Delta_R^{(p)}: R \to R \otimes R \otimes \cdots \otimes R, \tag{7.1.4}$$

where $R \otimes R \otimes \cdots \otimes R$ contains p factors, be the R-linear mapping in which $1 \mapsto 1 \otimes 1 \otimes \cdots \otimes 1$.

Now, by Section (3.2), $A_1 \otimes A_2 \otimes \cdots \otimes A_p$ is an R-algebra. In fact its multiplication and unit mappings are given by

$$\mu_{A_1 \otimes \cdots \otimes A_p} = (\mu_1 \otimes \mu_2 \otimes \cdots \otimes \mu_p) \circ \Lambda_p \tag{7.1.5}$$

and

$$\eta_{A_1 \otimes \cdots \otimes A_p} = (\eta_1 \otimes \eta_2 \otimes \cdots \otimes \eta_p) \circ \Delta_R^{(p)} \tag{7.1.6}$$

respectively.

We turn next to matters involving gradings. Suppose that (A, μ, η) is an R-algebra and let $\{A_n\}_{n \in \mathbb{Z}}$ be a family of R-submodules of A such that

$$A = \sum_{n \in \mathbb{Z}} A_n \quad \text{(d.s.)}, \tag{7.1.7}$$

that is we assume that $\{A_n\}_{n \in \mathbb{Z}}$ grades A *as an R-module*. Then this gives rise to a grading $\{(A \otimes A)_n\}_{n \in \mathbb{Z}}$ on the module $A \otimes A$, where

$$(A \otimes A)_n = \sum_{p+q=n} A_p \otimes A_q \quad \text{(d.s.)}. \tag{7.1.8}$$

(Of course (7.1.8) is the usual *total grading* on $A \otimes A$.) This said, if $\{A_n\}_{n \in \mathbb{Z}}$ happens to be an *algebra-grading*, then the mappings $\mu: A \otimes A \to A$ and $\eta: R \to A$ preserve degrees, it being understood that R is to be graded trivially. Conversely, if $\mu: A \otimes A \to A$ and $\eta: R \to A$ are degree-preserving, then $\{A_n\}_{n \in \mathbb{Z}}$ is an algebra-grading on A.

Suppose now that $A^{(1)}, A^{(2)}, \ldots, A^{(p)}$ are graded R-algebras and let $A^{(i)}$ have μ_i and η_i as its multiplication and unit mappings. Further, let $I = (i_1, i_2, \ldots, i_p)$ and $J = (j_1, j_2, \ldots, j_p)$ be sequences of p integers and put

$$N(I, J) = \sum_{r>s} i_r j_s, \tag{7.1.9}$$

$$e(I, J) = (-1)^{N(I,J)} \tag{7.1.10}$$

(see (3.4.2) and (3.4.3)). Then there is an isomorphism of

$$(A_{i_1}^{(1)} \otimes A_{i_2}^{(2)} \otimes \cdots \otimes A_{i_p}^{(p)}) \otimes (A_{j_1}^{(1)} \otimes A_{j_2}^{(2)} \otimes \cdots \otimes A_{j_p}^{(p)}) \tag{7.1.11}$$

onto

$$(A_{i_1}^{(1)} \otimes A_{j_1}^{(1)}) \otimes (A_{i_2}^{(2)} \otimes A_{j_2}^{(2)}) \otimes \cdots \otimes (A_{i_p}^{(p)} \otimes A_{j_p}^{(p)}) \tag{7.1.12}$$

in which, with a self-explanatory notation,

$$(a_{i_1} \otimes a_{i_2} \otimes \cdots \otimes a_{i_p}) \otimes (a'_{j_1} \otimes a'_{j_2} \otimes \cdots \otimes a'_{j_p})$$

is mapped into

$$e(I, J)(a_{i_1} \otimes a'_{j_1}) \otimes (a_{i_2} \otimes a'_{j_2}) \otimes \cdots \otimes (a_{i_p} \otimes a'_{j_p}).$$

But $(A^{(1)} \otimes A^{(2)} \otimes \cdots \otimes A^{(p)}) \otimes (A^{(1)} \otimes A^{(2)} \otimes \cdots \otimes A^{(p)})$ is the direct sum of the modules (7.1.11) whereas on the other hand $(A^{(1)} \otimes A^{(1)}) \otimes (A^{(2)} \otimes A^{(2)}) \otimes \cdots \otimes (A^{(p)} \otimes A^{(p)})$ is the direct sum of the modules (7.1.12). Consequently we can combine our various isomorphisms to obtain an isomorphism

$$\underline{\Lambda}_p\colon (A^{(1)} \otimes \cdots \otimes A^{(p)}) \otimes (A^{(1)} \otimes \cdots \otimes A^{(p)})$$
$$\xrightarrow{\sim} (A^{(1)} \otimes A^{(1)}) \otimes \cdots \otimes (A^{(p)} \otimes A^{(p)}). \tag{7.1.13}$$

We recall, from Section (3.4), that the modified tensor product $A^{(1)} \underline{\otimes} A^{(2)} \underline{\otimes} \cdots \underline{\otimes} A^{(p)}$ is a graded R-algebra. The multiplication and unit mappings of this algebra are given by

$$\mu_{A^{(1)} \underline{\otimes} \cdots \underline{\otimes} A^{(p)}} = (\mu_1 \otimes \mu_2 \otimes \cdots \otimes \mu_p) \circ \underline{\Lambda}_p \tag{7.1.14}$$

and

$$\eta_{A^{(1)} \underline{\otimes} \cdots \underline{\otimes} A^{(p)}} = (\eta_1 \otimes \eta_2 \otimes \cdots \otimes \eta_p) \circ \Delta_R^{(p)} \tag{7.1.15}$$

respectively, where $\Delta_R^{(p)}$ is the homomorphism previously encountered in (7.1.4).

Now that we have described some of the basic features of algebras in terms of mappings (rather than in terms of elements), we are ready to introduce the dual theory which was mentioned briefly in the introduction to this chapter.

7.2 Coalgebras

Let A be an R-module and this time suppose that we are given R-linear mappings $\Delta\colon A \to A \otimes A$ and $\varepsilon\colon A \to R$. The triplet (A, Δ, ε) is called an *R-coalgebra* provided that the diagrams

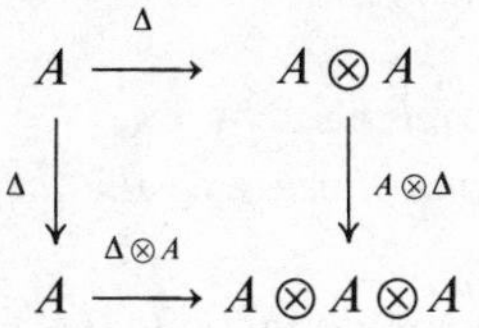

$$\begin{array}{ccc} A & \xrightarrow{\Delta} & A \otimes A \\ {\scriptstyle\Delta}\downarrow & & \downarrow{\scriptstyle A \otimes \Delta} \\ A & \xrightarrow{\Delta \otimes A} & A \otimes A \otimes A \end{array} \tag{7.2.1}$$

and

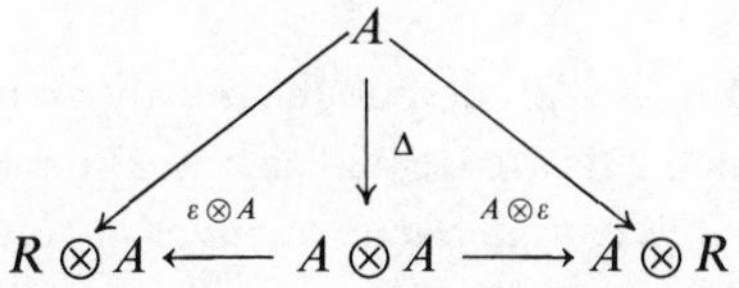

$$\begin{array}{ccccc} & \swarrow & A & \searrow & \\ & & \downarrow{\scriptstyle\Delta} & & \\ R \otimes A & \xleftarrow{\varepsilon \otimes A} & A \otimes A & \xrightarrow{A \otimes \varepsilon} & A \otimes R \end{array} \tag{7.2.2}$$

are commutative. Here $A \to A \otimes R$ maps a into $a \otimes 1$ and correspondingly for $A \to R \otimes A$.

If (A, Δ, ε) is a coalgebra, then $\Delta\colon A \to A \otimes A$ is called the *comultiplication* or *diagonalization* mapping of the coalgebra, and the fact that (7.2.1) is commutative is described by saying that comultiplication is *coassociative.* The mapping $\varepsilon\colon A \to R$ is known as the *counit*.

It is important to note that R itself becomes a coalgebra if we take $\Delta\colon R \to R \otimes R$ to be the R-linear mapping which carries 1 into $1 \otimes 1$ and we take $\varepsilon: R \to R$ to be the identity mapping. Whenever we speak of R as being a coalgebra it is always this structure that we have in mind.

Now suppose that $(A, \Delta_A, \varepsilon_A)$ and $(B, \Delta_B, \varepsilon_B)$ are coalgebras. A mapping $f: A \to B$ is called a *homomorphism of coalgebras* or a *coalgebra-homomorphism* provided that (i) f is R-linear, (ii) $\Delta_B \circ f = (f \otimes f) \circ \Delta_A$ and (iii) $\varepsilon_B \circ f = \varepsilon_A$. (We can describe conditions (ii) and (iii) by saying that f has to be compatible with comultiplication and it must preserve counits.) If a homomorphism of coalgebras is also a bijection, then it is called an *isomorphism of coalgebras* and when this is the case the inverse mapping is also a coalgebra-isomorphism. In the last paragraph we observed that R itself is a coalgebra. Now if Δ_R is the comultiplication of R and ε_R is its counit, then $\varepsilon_R \circ \varepsilon_A = \varepsilon_A$ and $\Delta_R \circ \varepsilon_A = (\varepsilon_A \otimes \varepsilon_A) \circ \Delta_A$. (The latter follows from the commutative property of (7.2.2).) Accordingly *the counit mapping* $\varepsilon_A: A \to R$ *is a homomorphism of R-coalgebras.*

Finally, homomorphisms of coalgebras can be combined, that is to say that if $f: A \to B$ and $g: B \to C$ are coalgebra-homomorphisms, then $g \circ f$ is a homomorphism of the coalgebra A into the coalgebra C.

7.3 Graded coalgebras

Let (A, Δ, ε) be an R-coalgebra and let $\{A_n\}_{n \in \mathbb{Z}}$ be a grading on A considered, for the moment, simply as an R-module. Then there is induced on the module $A \otimes A$ the usual total grading in which $(A \otimes A)_n$ is given by (7.1.8). Let R be given the trivial grading.

Definition. *We say that 'A is graded as a coalgebra by* $\{A_n\}_{n \in \mathbb{Z}}$*' or that '*$\{A_n\}_{n \in \mathbb{Z}}$ *is a coalgebra-grading on A' if* Δ *and* ε *preserve the degrees of homogeneous elements.*

Thus, in the case of a graded coalgebra, the counit ε maps all those homogeneous elements of A whose degree is different from zero into the zero element of R.

Any coalgebra can be turned into a graded coalgebra by giving it the trivial grading (i.e. all non-zero elements are treated as being homogeneous and of degree zero). *When R itself is considered as a graded coalgebra, it is always to be understood that the grading in question is the trivial one.*

Now let A and B be graded coalgebras. By a *homomorphism* $A \to B$ *of graded coalgebras* is meant a coalgebra-homomorphism which preserves degrees. Naturally if such a mapping happens to be a bijection, then it is called an *isomorphism of graded coalgebras* in which case its inverse will also be an isomorphism of graded coalgebras.

This is all that needs to be said at present on the subject of gradings. Our next consideration will be the question of how tensor products of coalgebras are to be formed.

7.4 Tensor products of coalgebras

Let $A_1, A_2, \ldots, A_p$ be R-modules. There is an isomorphism

$$V_p: (A_1 \otimes A_1) \otimes \cdots \otimes (A_p \otimes A_p) \xrightarrow{\sim} (A_1 \otimes \cdots \otimes A_p) \otimes (A_1 \otimes \cdots \otimes A_p) \tag{7.4.1}$$

in which, with a self-explanatory notation, the element $(a_1 \otimes a_1') \otimes \cdots \otimes (a_p \otimes a_p')$ of the first module is matched with the element $(a_1 \otimes \cdots \otimes a_p) \otimes (a_1' \otimes \cdots \otimes a_p')$ of the second. Thus V_p is an isomorphism which accomplishes a certain type of rearrangement. In like manner we shall use

$$W_p: (A_1 \otimes A_1 \otimes A_1) \otimes \cdots \otimes (A_p \otimes A_p \otimes A_p) \xrightarrow{\sim} (A_1 \otimes \cdots \otimes A_p) \otimes (A_1 \otimes \cdots \otimes A_p) \otimes (A_1 \otimes \cdots \otimes A_p) \tag{7.4.2}$$

for the rearranging isomorphism which makes

$$(a_1 \otimes a_1' \otimes a_1'') \otimes (a_2 \otimes a_2' \otimes a_2'') \otimes \cdots \otimes (a_p \otimes a_p' \otimes a_p'')$$

and

$$(a_1 \otimes a_2 \otimes \cdots \otimes a_p) \otimes (a_1' \otimes a_2' \otimes \cdots \otimes a_p') \otimes (a_1'' \otimes a_2'' \otimes \cdots \otimes a_p'')$$

correspond.

Now suppose that, for $1 \leq i \leq p$, $(A_i, \Delta_i, \varepsilon_i)$ is an R-coalgebra (not necessarily graded) and put

$$A = A_1 \otimes A_2 \otimes \cdots \otimes A_p. \tag{7.4.3}$$

At this stage A is just an R-module. We note that the mapping $\Delta_1 \otimes \Delta_2 \otimes \cdots \otimes \Delta_p$ has domain $A_1 \otimes A_2 \otimes \cdots \otimes A_p$ and codomain $(A_1 \otimes A_1) \otimes (A_2 \otimes A_2) \otimes \cdots \otimes (A_p \otimes A_p)$. Consequently, if we put

$$\Delta = V_p \circ (\Delta_1 \otimes \Delta_2 \otimes \cdots \otimes \Delta_p), \tag{7.4.4}$$

then Δ is an R-linear mapping of A into $A \otimes A$.

Consider the product $R \otimes R \otimes \cdots \otimes R$ where the number of factors is p. There is a homomorphism

$$\mu_R^{(p)}: R \otimes R \otimes \cdots \otimes R \to R \tag{7.4.5}$$

which satisfies $\mu_R^{(p)}(r_1 \otimes r_2 \otimes \cdots \otimes r_p) = r_1 r_2 \ldots r_p$; and using this we may define

$$\varepsilon: A \to R \tag{7.4.6}$$

by

$$\varepsilon = \mu_R^{(p)} \circ (\varepsilon_1 \otimes \varepsilon_2 \otimes \cdots \otimes \varepsilon_p). \tag{7.4.7}$$

Of course, ε is R-linear.

Theorem 1. *For $1 \leq i \leq p$ let $(A_i, \Delta_i, \varepsilon_i)$ be an R-coalgebra. Then with the above notation (A, Δ, ε) is also an R-coalgebra.*

Proof. The mapping

$$(A_1 \otimes \Delta_1) \circ \Delta_1 \otimes (A_2 \otimes \Delta_2) \circ \Delta_2 \otimes \cdots \otimes (A_p \otimes \Delta_p) \circ \Delta_p$$

takes $A_1 \otimes A_2 \otimes \cdots \otimes A_p$ into $(A_1 \otimes A_1 \otimes A_1) \otimes \cdots \otimes (A_p \otimes A_p \otimes A_p)$. Consequently

$$W_p \circ ((A_1 \otimes \Delta_1) \circ \Delta_1 \otimes \cdots \otimes (A_p \otimes \Delta_p) \circ \Delta_p) \tag{7.4.8}$$

maps A into $A \otimes A \otimes A$. *We claim that* (7.4.8) *is the same as* $(A \otimes \Delta) \circ \Delta$.

In order to verify this claim we shall introduce some notation which is rather cumbersome but which is fully explicit. Later, when dealing with similar but sometimes more complicated situations, we shall simplify the notation in order to prevent it becoming unwieldy. The following argument will provide a model for expanding the simplified expressions if the reader wishes to have more details.

Suppose that $a_1, a_2, \ldots, a_p$ belong to $A_1, A_2, \ldots, A_p$ respectively and let us write

$$\Delta_{A_1}(a_1) = \sum_{\lambda} a_1'(\lambda) \otimes a_1''(\lambda),$$

$$\Delta_{A_2}(a_2) = \sum_{\mu} a_2'(\mu) \otimes a_2''(\mu),$$

and so on until finally we have

$$\Delta_{A_p}(a_p) = \sum_{\tau} a_p'(\tau) \otimes a_p''(\tau).$$

(The simplification that we shall use later consists in replacing these relations by $\Delta_{A_i}(a_i) = \sum a_i' \otimes a_i''$ for $1 = 1, 2, \ldots, p$. Thus there will be no explicit references to the parameters $\lambda, \mu, \ldots, \tau$ used to index the terms in the various sums. The reader is to take their presence as understood.) It then follows that

$$\begin{aligned}(\Delta_1 \otimes \Delta_2 \otimes \cdots \otimes \Delta_p)(a_1 \otimes a_2 \otimes \cdots \otimes a_p) \\ = \left(\sum_{\lambda} a_1'(\lambda) \otimes a_1''(\lambda)\right) \otimes \left(\sum_{\mu} a_2'(\mu) \otimes a_2''(\mu)\right) \\ \otimes \cdots \otimes \left(\sum_{\tau} a_p'(\tau) \otimes a_p''(\tau)\right)\end{aligned}$$

whence

$$\begin{aligned}\Delta(a_1 \otimes a_2 \otimes \cdots \otimes a_p) \\ = \sum_{\lambda, \mu, \ldots, \tau} (a_1'(\lambda) \otimes a_2'(\mu) \otimes \cdots \otimes a_p'(\tau)) \otimes (a_1''(\lambda) \otimes a_2''(\mu) \\ \otimes \cdots \otimes a_p''(\tau))\end{aligned}$$

and therefore

$$(A \otimes \Delta)(\Delta(a_1 \otimes a_2 \otimes \cdots \otimes a_p))$$
$$= \sum_{\lambda,\mu,\ldots,\tau} (a_1'(\lambda) \otimes a_2'(\mu) \otimes \cdots \otimes a_p'(\tau)) \otimes \Delta(a_1''(\lambda) \otimes a_2''(\mu) \otimes \cdots \otimes a_p''(\tau)).$$

On the other hand,

$$((A_1 \otimes \Delta_1) \circ \Delta_1 \otimes \cdots \otimes (A_p \otimes \Delta_p) \circ \Delta_p)(a_1 \otimes a_2 \otimes \cdots \otimes a_p)$$
$$= \left(\sum_\lambda a_1'(\lambda) \otimes \Delta_1 a_1''(\lambda)\right) \otimes \cdots \otimes \left(\sum_\tau a_p'(\tau) \otimes \Delta_p a_p''(\tau)\right)$$

from which it follows that

$$W_p \circ ((A_1 \otimes \Delta_1) \circ \Delta_1 \otimes \cdots \otimes (A_p \otimes \Delta_p) \circ \Delta_p)$$

applied to $a_1 \otimes a_2 \otimes \cdots \otimes a_p$ yields

$$\sum_{\lambda,\mu,\ldots,\tau} (a_1'(\lambda) \otimes a_2'(\mu) \otimes \cdots \otimes a_p'(\tau)) \otimes V_p(\Delta_1 a_1''(\lambda) \otimes \Delta_2 a_2''(\mu) \otimes \cdots \otimes \Delta_p a_p''(\tau))$$
$$= \sum_{\lambda,\mu,\ldots,\tau} (a_1'(\lambda) \otimes a_2'(\mu) \otimes \cdots \otimes a_p'(\tau)) \otimes \Delta(a_1''(\lambda) \otimes a_2''(\mu) \otimes \cdots \otimes a_p''(\tau)).$$

This establishes our claim.

It has now been shown that

$$(A \otimes \Delta) \circ \Delta = W_p \circ ((A_1 \otimes \Delta_1) \circ \Delta_1 \otimes \cdots \otimes (A_p \otimes \Delta_p) \circ \Delta_p)$$

and similarly it can be proved that

$$(\Delta \otimes A) \circ \Delta = W_p \circ ((\Delta_1 \otimes A_1) \circ \Delta_1 \otimes \cdots \otimes (\Delta_p \otimes A_p) \circ \Delta_p).$$

But $(A_i \otimes \Delta_i) \circ \Delta_i = (\Delta_i \otimes A_i) \circ \Delta_i$ because Δ_i is coassociative. Accordingly $(A \otimes \Delta) \circ \Delta = \Delta \circ (A \otimes \Delta)$ and therefore Δ is coassociative as well.

It remains for us to demonstrate that

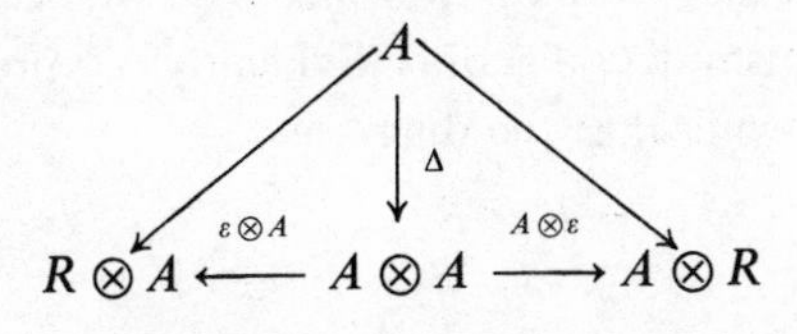

is a commutative diagram, and, as the two triangles can be treated similarly, we shall confine our attention to the one on the left-hand side. In this case it will suffice to show that the effect, on $a_1 \otimes a_2 \otimes \cdots \otimes a_p$, of applying in succession the homomorphisms

$$A \xrightarrow{\Delta} A \otimes A \xrightarrow{\varepsilon \otimes A} R \otimes A \xrightarrow{\sim} A$$

is to leave the element unchanged. Now, when these homomorphisms are applied the outcome is

$$\left(\sum_{\lambda} \varepsilon_1(a_1'(\lambda))a_1''(\lambda)\right) \otimes \cdots \otimes \left(\sum_{\tau} \varepsilon_p(a_p'(\tau))a_p''(\tau)\right).$$

But $\sum_\lambda \varepsilon_1(a_1'(\lambda))a_1''(\lambda) = a_1$ because the diagram

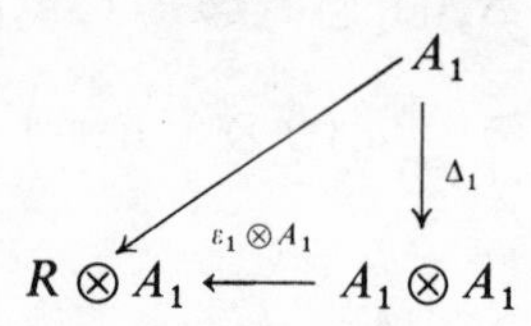

commutes, and similarly in the other cases. Accordingly the image of $a_1 \otimes a_2 \otimes \cdots \otimes a_p$ is itself and with this observation the proof is complete.

Definition. *The coalgebra* (A, Δ, ε) *of Theorem 1 is called the tensor product of the coalgebras* $A_1, A_2, \ldots, A_p$ *and it is denoted by* $A_1 \otimes A_2 \otimes \cdots \otimes A_p$.

Since R is both an algebra and a coalgebra, the p-fold product $R \otimes R \otimes \cdots \otimes R$ is also an algebra and a coalgebra. Now by (7.1.4) and (7.4.5) we have R-linear mappings

$$\Delta_R^{(p)}: R \to R \otimes R \otimes \cdots \otimes R$$

and

$$\mu_R^{(p)}: R \otimes R \otimes \cdots \otimes R \to R;$$

in fact $\Delta_R^{(p)}$ is the unit mapping of the algebra $R \otimes R \otimes \cdots \otimes R$ whereas $\mu_R^{(p)}$ is the counit mapping of the coalgebra $R \otimes R \otimes \cdots \otimes R$. It follows that $\Delta_R^{(p)}$ is an algebra-homomorphism and $\mu_R^{(p)}$ is a coalgebra-homomorphism. However, more is true as is shown by

Lemma 1. $\Delta_R^{(p)}$ *is a homomorphism of both R-algebras and R-coalgebras; and the same is true of* $\mu_R^{(p)}$.

Proof. It is clear that $\mu_R^{(p)}$ is an algebra-homomorphism so we need only show that $\Delta_R^{(p)}$ is a homomorphism of coalgebras. Evidently $\Delta_R^{(p)}$ preserves counits so we are left with showing that the diagram

$$\begin{array}{ccc} R & \longrightarrow & R \otimes R \\ \downarrow{\scriptstyle \Delta_R^{(p)}} & & \downarrow{\scriptstyle \Delta_R^{(p)} \otimes \Delta_R^{(p)}} \\ R \otimes \cdots \otimes R & \longrightarrow & (R \otimes \cdots \otimes R) \otimes (R \otimes \cdots \otimes R) \end{array}$$

commutes, where the horizontal mappings are comultiplications. However, this becomes clear as soon as we examine what happens to the identity element of R when it is transported to the bottom right-hand corner by the two available routes.

Now that tensor products of coalgebras have been defined it is natural to ask whether our results on tensor products of algebras have valid analogues. And in fact they do. However, partly because a coalgebra is a relatively unfamiliar object, and partly because of the greater use of mappings made in characterizing coalgebras, it is not an entirely trivial matter to adapt the arguments originally given in Chapter 3. Consequently, until the reader has had a chance to become familiar with the way in which adjustments may be made, we shall expound the dual theory in a fairly leisurely manner.

First suppose that $A_1, A_2, \ldots, A_p$ and $B_1, B_2, \ldots, B_p$ are R-coalgebras and let

$$f_i\colon A_i \to B_i \tag{7.4.9}$$

be a coalgebra-homomorphism for $i=1,2,\ldots,p$. Then certainly $f_1 \otimes f_2 \otimes \cdots \otimes f_p$ is a homomorphism of the *module* $A_1 \otimes A_2 \otimes \cdots \otimes A_p$ into the *module* $B_1 \otimes B_2 \otimes \cdots \otimes B_p$. However, as the following lemma shows, the homomorphism respects the coalgebra structures.

Lemma 2. *With the above assumptions*

$$f_1 \otimes \cdots \otimes f_p\colon A_1 \otimes A_2 \otimes \cdots \otimes A_p \to B_1 \otimes B_2 \otimes \cdots \otimes B_p$$

is a homomorphism of coalgebras.

Proof. Put $f=f_1 \otimes f_2 \otimes \cdots \otimes f_p$, $A=A_1 \otimes A_2 \otimes \cdots \otimes A_p$, $B=B_1 \otimes B_2 \otimes \cdots \otimes B_p$, and let a_i belong to A_i $(i=1,2,\ldots,p)$. If now, using the abbreviated notation (see the proof of Theorem 1), we set $\Delta_{A_i}(a_i)=\sum a_i' \otimes a_i''$, then

$$\Delta_A(a_1 \otimes a_2 \otimes \cdots \otimes a_p)=\sum (a_1' \otimes \cdots \otimes a_p') \otimes (a_1'' \otimes \cdots \otimes a_p'')$$

and hence

$$\begin{aligned}&(f \otimes f)(\Delta_A(a_1 \otimes \cdots \otimes a_p))\\ &=\sum (f_1a_1' \otimes \cdots \otimes f_pa_p') \otimes (f_1a_1'' \otimes \cdots \otimes f_pa_p'').\end{aligned}$$

Again, because f_i is a homomorphism of coalgebras,

$$\Delta_{B_i}(f_ia_i)=\sum f_ia_i' \otimes f_ia_i''$$

and therefore

$$\begin{aligned}&\Delta_B(f_1a_1 \otimes \cdots \otimes f_pa_p)\\ &=\sum (f_1a_1' \otimes \cdots \otimes f_pa_p') \otimes (f_1a_1'' \otimes \cdots \otimes f_pa_p'').\end{aligned}$$

This shows that $\Delta_B \circ f=(f \otimes f) \circ \Delta_A$.

In the case of the counits we have

$$(\varepsilon_B \circ f)(a_1 \otimes \cdots \otimes a_p) = \varepsilon_{B_1}(f_1 a_1)\varepsilon_{B_2}(f_2 a_2) \ldots \varepsilon_{B_p}(f_p a_p).$$

But $\varepsilon_{B_i}(f_i a_i) = \varepsilon_{A_i}(a_i)$. Consequently

$$\begin{aligned}(\varepsilon_B \circ f)(a_1 \otimes \cdots \otimes a_p) &= \varepsilon_{A_1}(a_1)\varepsilon_{A_2}(a_2) \ldots \varepsilon_{A_p}(a_p) \\ &= \varepsilon_A(a_1 \otimes \cdots \otimes a_p).\end{aligned}$$

Thus $\varepsilon_B \circ f = \varepsilon_A$ and the proof is complete.

We turn next to the basic isomorphism theorems for coalgebras and prove results which are the counterparts of Theorems 2, 3 and 4 of Chapter 3.

Theorem 2. *Let $A_1, A_2, \ldots, A_p$ and $B_1, B_2, \ldots, B_q$ be R-coalgebras. Then the canonical isomorphism*

$$\begin{aligned}&A_1 \otimes \cdots \otimes A_p \otimes B_1 \otimes \cdots \otimes B_q \\ &\approx (A_1 \otimes \cdots \otimes A_p) \otimes (B_1 \otimes \cdots \otimes B_q)\end{aligned}$$

of R-modules (see Chapter 2, Theorem 1) is actually an isomorphism of R-coalgebras.

Proof. Put $A = A_1 \otimes \cdots \otimes A_p$, $B = B_1 \otimes \cdots \otimes B_q$ and $C = A_1 \otimes \cdots \otimes A_p \otimes B_1 \otimes \cdots \otimes B_q$. In what follows a_i will denote an element of A_i and b_j an element of B_j; furthermore we shall suppose that $\Delta_{A_i}(a_i) = \sum a_i' \otimes a_i''$ and $\Delta_{B_j}(b_j) = \sum b_j' \otimes b_j''$. Lastly we shall use

$$\begin{aligned}&\phi\colon A_1 \otimes \cdots \otimes A_p \otimes B_1 \otimes \cdots \otimes B_q \\ &\xrightarrow{\sim} (A_1 \otimes \cdots \otimes A_p) \otimes (B_1 \otimes \cdots \otimes B_q)\end{aligned}$$

to denote the canonical module-isomorphism. Of course, it will suffice to show that ϕ is a homomorphism of coalgebras.

To this end we first observe that the result of applying Δ_C to $a_1 \otimes \cdots \otimes a_p \otimes b_1 \otimes \cdots \otimes b_q$ is

$$\sum (a_1' \otimes \cdots \otimes a_p' \otimes b_1' \otimes \cdots \otimes b_q') \otimes (a_1'' \otimes \cdots \otimes a_p'' \otimes b_1'' \otimes \cdots \otimes b_q'')$$

and therefore when $(\phi \otimes \phi) \circ \Delta_C$ is applied to the same element what we obtain is

$$\sum [(a_1' \otimes \cdots \otimes a_p') \otimes (b_1' \otimes \cdots \otimes b_q')] \otimes [(a_1'' \otimes \cdots \otimes a_p'') \otimes (b_1'' \otimes \cdots \otimes b_q'')]. \tag{7.4.10}$$

But

$$\begin{aligned}&\phi(a_1 \otimes \cdots \otimes a_p \otimes b_1 \otimes \cdots \otimes b_q) \\ &\quad = (a_1 \otimes \cdots \otimes a_p) \otimes (b_1 \otimes \cdots \otimes b_q) = a \otimes b\end{aligned}$$

say, and we also have

$$\Delta_A(a) = \sum (a_1' \otimes \cdots \otimes a_p') \otimes (a_1'' \otimes \cdots \otimes a_p'')$$

and

$$\Delta_B(b) = \sum (b_1' \otimes \cdots \otimes b_q') \otimes (b_1'' \otimes \cdots \otimes b_q'').$$

Accordingly $\Delta_{A \otimes B} \circ \phi$ applied to $a_1 \otimes \cdots \otimes a_p \otimes b_1 \otimes \cdots \otimes b_q$ produces the element (7.4.10) and so we have established that $(\phi \otimes \phi) \circ \Delta_C = \Delta_{A \otimes B} \circ \phi$.

Finally when $\varepsilon_{A \otimes B} \circ \phi$ and ε_C are made to act on $a_1 \otimes \cdots \otimes a_p \otimes b_1 \otimes \cdots \otimes b_q$ the result in both cases is

$$\varepsilon_{A_1}(a_1) \ldots \varepsilon_{A_p}(a_p) \varepsilon_{B_1}(b_1) \ldots \varepsilon_{B_q}(b_q).$$

It follows that $\varepsilon_{A \otimes B} \circ \phi = \varepsilon_C$ and now the proof is complete.

Theorem 3. *Let $A_1, A_2, \ldots, A_p$ be R-coalgebras and let $(i_1, i_2, \ldots, i_p)$ be a permutation of $(1, 2, \ldots, p)$. Then the canonical isomorphism*

$$A_1 \otimes A_2 \otimes \cdots \otimes A_p \approx A_{i_1} \otimes A_{i_2} \otimes \cdots \otimes A_{i_p}$$

of R-modules (see Chapter 2, Theorem 2) is in fact an isomorphism of R-coalgebras.

Proof. Let $\phi: A_1 \otimes A_2 \otimes \cdots \otimes A_p \xrightarrow{\sim} A_{i_1} \otimes A_{i_2} \otimes \cdots \otimes A_{i_p}$ be the module-isomorphism in question. For $1 \leq i \leq p$ let a_i belong to A_i and suppose that $\Delta_{A_i}(a_i) = \sum a_i' \otimes a_i''$. Put $A = A_1 \otimes A_2 \otimes \cdots \otimes A_p$ and $B = A_{i_1} \otimes A_{i_2} \otimes \cdots \otimes A_{i_p}$. If now $(\phi \otimes \phi) \circ \Delta_A$ and $\Delta_B \circ \phi$ are applied to $a_1 \otimes a_2 \otimes \cdots \otimes a_p$, then the result in both cases is

$$\sum (a_{i_1}' \otimes \cdots \otimes a_{i_p}') \otimes (a_{i_1}'' \otimes \cdots \otimes a_{i_p}''),$$

whereas if $\varepsilon_B \circ \phi$ and ε_A are made to operate on the same element, then the result, again in both cases, is $\varepsilon_{A_1}(a_1)\varepsilon_{A_2}(a_2) \ldots \varepsilon_{A_p}(a_p)$. The theorem follows.

We recall that R itself is a coalgebra.

Theorem 4. *Let A be an R-coalgebra. Then the canonical module-isomorphisms $A \otimes R \approx A$ and $R \otimes A \approx A$ are isomorphisms of R-coalgebras.*

Proof. Let $\phi: A \otimes R \to A$ be the module-isomorphism in which $\phi(a \otimes r) = ra$, let α belong to A, and let $\Delta_A(\alpha) = \sum \alpha' \otimes \alpha''$. Then

$$(\phi \otimes \phi)(\Delta_{A \otimes R}(\alpha \otimes r)) = r \sum \alpha' \otimes a'' = \Delta_A(\phi(\alpha \otimes r))$$

so that $(\phi \otimes \phi) \circ \Delta_{A \otimes R} = \Delta_A \circ \phi$. Again

$$\varepsilon_A(\phi(\alpha \otimes r)) = r\varepsilon_A(\alpha) = \varepsilon_{A \otimes R}(\alpha \otimes r)$$

and therefore $\varepsilon_A \circ \phi = \varepsilon_{A \otimes R}$. This shows that $A \otimes R \approx A$ is a coalgebra-isomorphism and the other isomorphism is dealt with similarly.

We shall now examine, fairly briefly, how this theory may be refined when we are dealing with graded coalgebras. To this end let $A^{(1)}, A^{(2)}, \ldots, A^{(p)}$ be *graded* R-coalgebras. Then, as we have seen,

$$A = A^{(1)} \otimes A^{(2)} \otimes \cdots \otimes A^{(p)} \tag{7.4.11}$$

has a natural structure as a coalgebra. In what follows Δ_A and ε_A will denote the comultiplication and the counit of this coalgebra.

Let $I = (i_1, i_2, \ldots, i_p)$ be a sequence of p integers, and (as in the case of graded algebras) put

$$A_I = A_{i_1}^{(1)} \otimes A_{i_2}^{(2)} \otimes \cdots \otimes A_{i_p}^{(p)} \tag{7.4.12}$$

so that

$$A = \sum_I A_I \tag{7.4.13}$$

this being a direct sum of R-modules. If we now set

$$|I| = i_1 + i_2 + \cdots + i_p \tag{7.4.14}$$

and

$$A_n = \sum_{|I|=n} A_I, \tag{7.4.15}$$

where the sum is taken in A, then $\{A_n\}_{n \in \mathbb{Z}}$ is a grading on the module A. It is clear that if R is graded trivially, then $\varepsilon_A: A \to R$ preserves the degrees of homogeneous elements; and it follows from (7.4.4) that $\Delta_A: A \to A \otimes A$ is also degree-preserving provided that $A \otimes A$ is given the usual total grading.

We can sum up these observations as follows. Let $A^{(1)}, A^{(2)}, \ldots, A^{(p)}$ be graded coalgebras. *Then* $A^{(1)} \otimes A^{(2)} \otimes \cdots \otimes A^{(p)}$ *has the structure of a coalgebra by virtue of Theorem 1. Furthermore the usual total grading on the module* $A^{(1)} \otimes A^{(2)} \otimes \cdots \otimes A^{(p)}$ *is a coalgebra-grading.*

In the next section we shall discuss a way of modifying this structure. For the present, however, we shall leave this refinement on one side and begin by observing that if $A^{(1)}, A^{(2)}, \ldots, A^{(p)}$ and $B^{(1)}, B^{(2)}, \ldots, B^{(q)}$ are graded coalgebras, then the isomorphism

$$\begin{aligned} &A^{(1)} \otimes \cdots \otimes A^{(p)} \otimes B^{(1)} \otimes \cdots \otimes B^{(q)} \\ &\qquad \approx (A^{(1)} \otimes \cdots \otimes A^{(p)}) \otimes (B^{(1)} \otimes \cdots \otimes B^{(q)}), \end{aligned} \tag{7.4.16}$$

provided by Theorem 2, is an isomorphism of *graded* coalgebras; and that if $(i_1, i_2, \ldots, i_p)$ is a permutation of $(1, 2, \ldots, p)$, then the isomorphism

$$A^{(1)} \otimes A^{(2)} \otimes \cdots \otimes A^{(p)} \approx A^{(i_1)} \otimes A^{(i_2)} \otimes \cdots \otimes A^{(i_p)} \tag{7.4.17}$$

of Theorem 3 is also an isomorphism of *graded* coalgebras. Yet again, if A is a graded coalgebra, then the coalgebra-isomorphisms

$$A \otimes R \approx A \tag{7.4.18}$$

and

$$R \otimes A \approx A \tag{7.4.19}$$

of Theorem 4 preserve degrees provided that R is graded trivially. Thus (7.4.18) and (7.4.19) are isomorphisms of *graded* coalgebras.

Finally, suppose that

$$f_i: A^{(i)} \to B^{(i)}$$

is a homomorphism of graded coalgebras for $i=1,2,\ldots,p$. Then, because

$$f_1 \otimes \cdots \otimes f_p: A^{(1)} \otimes \cdots \otimes A^{(p)} \to B^{(1)} \otimes \cdots \otimes B^{(p)}$$

preserves degrees, it is a homomorphism of *graded* coalgebras on account of Lemma 2.

7.5 Modified tensor products of coalgebras

In this section we shall develop ideas that correspond to those described in Section (3.4). Suppose then that $A^{(1)}, A^{(2)}, \ldots, A^{(p)}$ are graded R-coalgebras. We have already seen how the R-module

$$A = A^{(1)} \otimes A^{(2)} \otimes \cdots \otimes A^{(p)} \tag{7.5.1}$$

can be turned into a graded coalgebra $(A, \Delta_A, \varepsilon_A)$. We now propose to modify the coalgebra structure on A. The new coalgebra will have the same R-module structure, the same grading, and the same counit as before; only the comultiplication will be changed and this in a comparatively simple way. We are, as it were, putting a twist into (7.5.1).

Let $I=(i_1, i_2, \ldots, i_p)$ and $J=(j_1, j_2, \ldots, j_p)$ be sequences of p integers, and, as in (7.1.9) and (7.1.10), put

$$N(I,J) = \sum_{r>s} i_r j_s \tag{7.5.2}$$

and

$$e(I,J) = (-1)^{N(I,J)}. \tag{7.5.3}$$

We shall use $I+J$ to denote the sequence $(i_1+j_1, i_2+j_2, \ldots, i_p+j_p)$.

For given I and J, there is an isomorphism of

$$(A_{i_1}^{(1)} \otimes A_{j_1}^{(1)}) \otimes (A_{i_2}^{(2)} \otimes A_{j_2}^{(2)}) \otimes \cdots \otimes (A_{i_p}^{(p)} \otimes A_{j_p}^{(p)}) \tag{7.5.4}$$

onto

$$(A_{i_1}^{(1)} \otimes A_{i_2}^{(2)} \otimes \cdots \otimes A_{i_p}^{(p)}) \otimes (A_{j_1}^{(1)} \otimes A_{j_2}^{(2)} \otimes \cdots \otimes A_{j_p}^{(p)}) \tag{7.5.5}$$

in which

$$(a_{i_1} \otimes a'_{j_1}) \otimes (a_{i_2} \otimes a'_{j_2}) \otimes \cdots \otimes (a_{i_p} \otimes a'_{j_p})$$

is mapped into

$$e(I,J)(a_{i_1} \otimes a_{i_2} \otimes \cdots \otimes a_{i_p}) \otimes (a'_{j_1} \otimes a'_{j_2} \otimes \cdots \otimes a'_{j_p})$$

and these various isomorphisms can be combined to give an isomorphism

$$\begin{aligned} \underline{V}_p: (A^{(1)} \otimes A^{(1)}) \otimes \cdots \otimes (A^{(p)} \otimes A^{(p)}) \\ \xrightarrow{\sim} (A^{(1)} \otimes \cdots \otimes A^{(p)}) \otimes (A^{(1)} \otimes \cdots \otimes A^{(p)}) \end{aligned} \tag{7.5.6}$$

of R-modules.

Let us compare $\underline{V}_p$ with the untwisted isomorphism

$$V_p: (A^{(1)} \otimes A^{(1)}) \otimes \cdots \otimes (A^{(p)} \otimes A^{(p)})$$

$$\xrightarrow{\sim} (A^{(1)} \otimes \cdots \otimes A^{(p)}) \otimes (A^{(1)} \otimes \cdots \otimes A^{(p)})$$

which operates as in (7.4.1). If we denote by $\pi_{I,J}$ the projection of $(A^{(1)} \otimes \cdots \otimes A^{(p)}) \otimes (A^{(1)} \otimes \cdots \otimes A^{(p)})$ onto $A_I \otimes A_J$, that is onto the module (7.5.5), then we find that

$$\pi_{I,J} \circ \underline{V}_p = e(I, J)\pi_{I,J} \circ V_p. \tag{7.5.7}$$

We next define an R-linear, degree-preserving mapping $\underline{\Delta}: A \to A \otimes A$ by means of the formula

$$\underline{\Delta} = \underline{V}_p \circ (\Delta_{A^{(1)}} \otimes \Delta_{A^{(2)}} \otimes \cdots \otimes \Delta_{A^{(p)}}). \tag{7.5.8}$$

Now, by Theorem 1, $A = A^{(1)} \otimes A^{(2)} \otimes \cdots \otimes A^{(p)}$ is a coalgebra. Let us use Δ and ε to denote its comultiplication and counit mappings. Then, by (7.4.4), (7.5.7) and (7.5.8),

$$\pi_{I,J} \circ \underline{\Delta} = e(I, J)\pi_{I,J} \circ \Delta. \tag{7.5.9}$$

Lemma 3. *The triple* $(A^{(1)} \otimes \cdots \otimes A^{(p)}, \underline{\Delta}, \varepsilon)$ *is a coalgebra and the usual total grading on* $A^{(1)} \otimes A^{(2)} \otimes \cdots \otimes A^{(p)}$ *grades the coalgebra.*

Proof. It will suffice to prove the first statement and for this a little temporary terminology will be useful. Let us say that an element α of A is *pure* if it has the form $\alpha_{i_1} \otimes \alpha_{i_2} \otimes \cdots \otimes \alpha_{i_p}$ (where $\alpha_{i_\nu} \in A_{i_\nu}^{(\nu)}$) and when this is the case let us use $\{\alpha\}$ to denote the sequence $(i_1, i_2, \ldots, i_p)$. Pure elements are, of course, homogeneous and every element of A is a finite sum of pure elements.

The idea is to use the fact that (A, Δ, ε) is already known to be a graded coalgebra. Suppose then that α belongs to A and is pure. Then $\Delta(\alpha)$ can be expressed in the form

$$\Delta(\alpha) = \sum \alpha' \otimes \alpha'',$$

where α', α'' are pure and $\{\alpha'\} + \{\alpha''\} = \{\alpha\}$. Note that $\pi_{I,J}(\Delta(\alpha)) = 0$ unless $\{\alpha\} = I + J$. Note, too, that

$$\underline{\Delta}(\alpha) = \sum e(\{\alpha'\}, \{\alpha''\})\alpha' \otimes \alpha''.$$

Now suppose that I, J, K are three sequences each consisting of p integers and let $\pi_{I,J,K}$ be the projection of $A \otimes A \otimes A$ onto $A_I \otimes A_J \otimes A_K$. It is clear that if we identify $A \otimes A \otimes A$ with $A \otimes (A \otimes A)$ in the usual way, then $\pi_{I,J,K} = \pi_I \otimes \pi_{J,K}$. Accordingly

$$(\pi_{I,J,K} \circ (A \otimes \Delta) \circ \Delta)(\alpha) = \sum \pi_I(\alpha') \otimes \pi_{J,K}(\Delta\alpha'')$$

and

$$(\pi_{I,J,K}\circ(A\otimes\underline{\Delta})\circ\underline{\Delta})(\alpha)$$
$$=\sum e(\{\alpha'\},\{\alpha''\})\pi_I(\alpha')\otimes\pi_{J,K}(\underline{\Delta}\alpha'').$$

But $\pi_{J,K}(\underline{\Delta}\alpha'')=e(J,K)\pi_{J,K}(\Delta\alpha'')$ by (7.5.9) and therefore

$$(\pi_{I,J,K}\circ(A\otimes\underline{\Delta})\circ\underline{\Delta})(\alpha)$$
$$=e(J,K)\sum e(\{\alpha'\},\{\alpha''\})\pi_I(\alpha')\otimes\pi_{J,K}(\Delta\alpha'')$$
$$=e(J,K)e(I,J+K)\sum\pi_I(\alpha')\otimes\pi_{J,K}(\Delta\alpha'')$$

because $\pi_I(\alpha')=0$ unless $\{\alpha'\}=I$, and $\pi_{J,K}(\Delta\alpha'')=0$ unless $\{\alpha''\}=J+K$. This shows that

$$(\pi_{I,J,K}\circ(A\otimes\underline{\Delta})\circ\underline{\Delta})(\alpha)=e(J,K)e(I,J+K)(\pi_{I,J,K}\circ(A\otimes\Delta)\circ\Delta)(\alpha)$$

and in a similar manner we can show that

$$(\pi_{I,J,K}\circ(\underline{\Delta}\otimes A)\circ\underline{\Delta})(\alpha)=e(I,J)e(I+J,K)(\pi_{I,J,K}\circ(\Delta\otimes A)\circ\Delta)(\alpha).$$

However, it is easy to verify that

$$e(I,J+K)e(J,K)=e(I,J)e(I+J,K)$$

and we know that $(A\otimes\Delta)\circ\Delta=(\Delta\otimes A)\circ\Delta$ because Δ is a comultiplication. Accordingly

$$\pi_{I,J,K}\circ(A\otimes\underline{\Delta})\circ\underline{\Delta}=\pi_{I,J,K}\circ(\underline{\Delta}\otimes A)\circ\underline{\Delta}$$

for all I, J, K and therefore $(A\otimes\underline{\Delta})\circ\underline{\Delta}=(\underline{\Delta}\otimes A)\circ\underline{\Delta}$.

It remains to be shown that the diagram

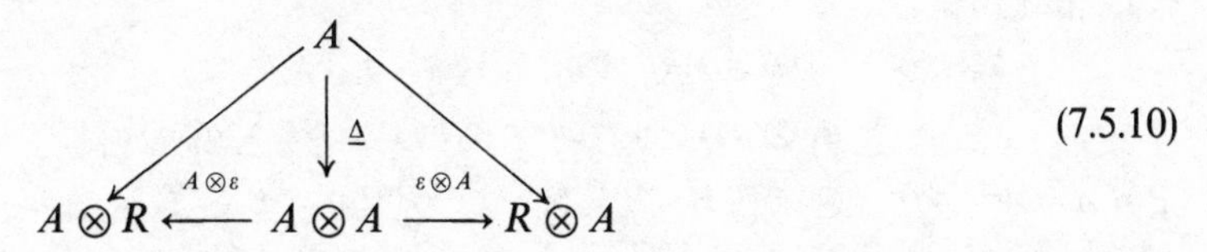

(7.5.10)

commutes. Now the combined effect, on α, of the mappings

$$A\xrightarrow{\underline{\Delta}}A\otimes A\xrightarrow{A\otimes\varepsilon}A\otimes R\xrightarrow{\sim}A$$

is to turn it into $\sum e(\{\alpha'\},\{\alpha''\})\varepsilon(\alpha'')\alpha'$. But if $\varepsilon(\alpha'')\neq 0$, then $\{\alpha''\}=(0,0,\ldots,0)$ and therefore $e(\{\alpha'\},\{\alpha''\})=1$. It follows that

$$\sum e(\{\alpha'\},\{\alpha''\})\varepsilon(\alpha'')\alpha'=\sum\varepsilon(\alpha'')\alpha'$$

and this is just α because the mappings

$$A\xrightarrow{\Delta}A\otimes A\xrightarrow{A\otimes\varepsilon}A\otimes R\xrightarrow{\sim}A$$

combine to give the identity mapping of A. This proves that the left-hand triangle in (7.5.10) commutes and the right-hand triangle can be dealt with in the same way. The proof is now complete.

The graded coalgebra $(A^{(1)} \otimes \cdots \otimes A^{(p)}, \underline{\Delta}, \varepsilon)$ will be called the *modified tensor product* or the *twisted tensor product* of $A^{(1)}, A^{(2)}, \ldots, A^{(p)}$; and it will be denoted by $A^{(1)} \underline{\otimes} A^{(2)} \underline{\otimes} \cdots \underline{\otimes} A^{(p)}$ to distinguish it from the coalgebra $A^{(1)} \otimes A^{(2)} \otimes \cdots \otimes A^{(p)}$.

We proceed to establish the basic properties of these modified tensor products. Suppose first that

$$f_i \colon A^{(i)} \to B^{(i)} \quad (i=1,2,\ldots,p) \tag{7.5.11}$$

is a homomorphism of graded coalgebras. Then $f_1 \otimes f_2 \otimes \cdots \otimes f_p$ is a homomorphism of the coalgebra $A^{(1)} \otimes A^{(2)} \otimes \cdots \otimes A^{(p)}$ into the coalgebra $B^{(1)} \otimes B^{(2)} \otimes \cdots \otimes B^{(p)}$. But changing to modified tensor products does not affect the underlying module structures. Consequently

$$f_1 \otimes \cdots \otimes f_p \colon A^{(1)} \underline{\otimes} \cdots \underline{\otimes} A^{(p)} \to B^{(1)} \underline{\otimes} \cdots \underline{\otimes} B^{(p)}. \tag{7.5.12}$$

Note that (7.5.12) is R-linear and preserves degrees.

Lemma 4. *If $f_1 \otimes f_2 \otimes \cdots \otimes f_p$ is considered as a mapping of $A^{(1)} \underline{\otimes} A^{(2)} \underline{\otimes} \cdots \underline{\otimes} A^{(p)}$ into $B^{(1)} \underline{\otimes} B^{(2)} \underline{\otimes} \cdots \underline{\otimes} B^{(p)}$, then it is a homomorphism of graded coalgebras.*

Proof. Consider the element $a_{i_1} \otimes a_{i_2} \otimes \cdots \otimes a_{i_p}$, where $a_{i_\nu} \in A^{(\nu)}_{i_\nu}$. We can write

$$\Delta_{A^{(\mu)}}(a_{i_\mu}) = \sum a'_\mu \otimes a''_\mu,$$

where a'_μ and a''_μ are homogeneous elements of $A^{(\mu)}$ the sum of whose degrees is i_μ, and then

$$\begin{aligned}(\Delta_{A^{(1)}} \otimes \cdots \otimes \Delta_{A^{(p)}})&(a_{i_1} \otimes a_{i_2} \otimes \cdots \otimes a_{i_p}) \\ &= \left(\sum a'_1 \otimes a''_1\right) \otimes \left(\sum a'_2 \otimes a''_2\right) \otimes \cdots \otimes \left(\sum a'_p \otimes a''_p\right).\end{aligned}$$

Put $A = A^{(1)} \underline{\otimes} \cdots \underline{\otimes} A^{(p)}$, $B = B^{(1)} \underline{\otimes} \cdots \underline{\otimes} B^{(p)}$, $f = f_1 \otimes f_2 \otimes \cdots \otimes f_p$ and set $a' = a'_1 \otimes \cdots \otimes a'_p$, $a'' = a''_1 \otimes \cdots \otimes a''_p$. If now we use $\{a'\}$ to denote the sequence formed by the degrees of $a'_1, a'_2, \ldots, a'_p$ and we define $\{a''\}$ similarly, then we find that

$$\Delta_A(a_{i_1} \otimes a_{i_2} \otimes \cdots \otimes a_{i_p}) = \sum e(\{a'\}, \{a''\}) a' \otimes a''.$$

Accordingly

$$((f \otimes f) \circ \Delta_A)(a_{i_1} \otimes \cdots \otimes a_{i_p}) = \sum e(\{a'\}, \{a''\}) f(a') \otimes f(a'').$$

But

$$\Delta_B(\mu)(f_\mu a_{i_\mu}) = \sum f_\mu a'_\mu \otimes f_\mu a''_\mu$$

and $\{fa'\} = \{a'\}$, $\{fa''\} = \{a''\}$ because f_μ is a homomorphism of the graded coalgebra $A^{(\mu)}$ into the graded coalgebra $B^{(\mu)}$. We therefore have

$$\begin{aligned}\Delta_B(f_1 a_{i_1} \otimes \cdots \otimes f_p a_{i_p}) &= \sum e(\{fa'\}, \{fa''\}) f(a') \otimes f(a'') \\ &= \sum e(\{a'\}, \{a''\}) f(a') \otimes f(a'') \\ &= ((f \otimes f) \circ \Delta_A)(a_{i_1} \otimes \cdots \otimes a_{i_p})\end{aligned}$$

from which it follows that $\Delta_B \circ f = (f \otimes f) \circ \Delta_A$. Thus f is compatible with the appropriate comultiplications.

We still have to satisfy ourselves that f preserves counits. However, this follows from the corresponding result for unmodified tensor products because the mappings involved are exactly the same in both cases.

The next result corresponds to Theorem 6 of Chapter 3. In order to state it we suppose that $A^{(1)}, A^{(2)}, \ldots, A^{(p)}$ and $B^{(1)}, B^{(2)}, \ldots, B^{(q)}$ are graded R-coalgebras. Then, by Theorem 1 of Chapter 2, there is an R-linear bijection

$$\begin{aligned}\phi\colon A^{(1)} \underline{\otimes} \cdots \underline{\otimes} A^{(p)} \underline{\otimes} B^{(1)} \underline{\otimes} \cdots \underline{\otimes} B^{(q)} \\ \xrightarrow{\sim} (A^{(1)} \underline{\otimes} \cdots \underline{\otimes} A^{(p)}) \underline{\otimes} (B^{(1)} \underline{\otimes} \cdots \underline{\otimes} B^{(q)})\end{aligned} \tag{7.5.13}$$

in which

$$\phi(a_{i_1} \otimes \cdots \otimes a_{i_p} \otimes b_{j_1} \otimes \cdots \otimes b_{j_q}) = (a_{i_1} \otimes \cdots \otimes a_{i_p}) \otimes (b_{j_1} \otimes \cdots \otimes b_{j_q}).$$

Theorem 5. *The bijection* (7.5.13) *is an isomorphism*

$$\begin{aligned}A^{(1)} \underline{\otimes} \cdots \underline{\otimes} A^{(p)} \underline{\otimes} B^{(1)} \underline{\otimes} \cdots \underline{\otimes} B^{(q)} \\ \xrightarrow{\sim} (A^{(1)} \underline{\otimes} \cdots \underline{\otimes} A^{(p)}) \underline{\otimes} (B^{(1)} \underline{\otimes} \cdots \underline{\otimes} B^{(q)})\end{aligned}$$

of graded coalgebras.

Proof. It is evident that ϕ preserves degrees and that it preserves counits. Consequently we need only show that it is compatible with the appropriate comultiplication mappings. In doing this we shall put

$$A = A^{(1)} \underline{\otimes} A^{(2)} \underline{\otimes} \cdots \underline{\otimes} A^{(p)},$$
$$B = B^{(1)} \underline{\otimes} B^{(2)} \underline{\otimes} \cdots \underline{\otimes} B^{(q)},$$

and

$$C = A^{(1)} \underline{\otimes} \cdots \underline{\otimes} A^{(p)} \underline{\otimes} B^{(1)} \underline{\otimes} \cdots \underline{\otimes} B^{(q)}$$

in order to simplify the notation.

Now suppose that $a_{i_\mu} \in A^{(\mu)}_{i_\mu}$ and $b_{j_\nu} \in B^{(\nu)}_{j_\nu}$. Then

$$\Delta_{A^{(\mu)}}(a_{i_\mu}) = \sum a'_\mu \otimes a''_\mu,$$
$$\Delta_{B^{(\nu)}}(b_{j_\nu}) = \sum b'_\nu \otimes b''_\nu,$$

where a'_μ, a''_μ are homogeneous elements of $A^{(\mu)}$ the sum of whose degrees is i_μ, and b'_ν, b''_ν are homogeneous elements of $B^{(\nu)}$ the sum of their degrees being j_ν. Since

$$\begin{aligned}(\Delta_{A^{(1)}} \otimes \cdots \otimes \Delta_{A^{(p)}} \otimes \Delta_{B^{(1)}} \otimes \cdots \otimes \Delta_{B^{(q)}})(a_{i_1} \otimes \cdots \otimes a_{i_p} \otimes b_{j_1} \otimes \cdots \otimes b_{j_q}) \\ = (\textstyle\sum a'_1 \otimes a''_1) \otimes \cdots \otimes \\ (\textstyle\sum a'_p \otimes a''_p) \otimes (\sum b'_1 \otimes b''_1) \otimes \cdots \otimes (\sum b'_q \otimes b''_q)\end{aligned}$$

it follows that

$$\Delta_C(a_{i_1} \otimes \cdots \otimes a_{i_p} \otimes b_{j_1} \otimes \cdots \otimes b_{j_q})$$
$$= \sum e(\{a'_1 \otimes \cdots \otimes b'_q\}, \{a''_1 \otimes \cdots \otimes b''_q\})(a'_1 \otimes \cdots \otimes b'_q) \otimes (a''_1 \otimes \cdots \otimes b''_q),$$

where by $\{a'_1 \otimes \cdots \otimes b'_q\}$ is meant the sequence formed by the degrees of $a'_1, \ldots, a'_p, b'_1, \ldots, b'_q$ taken in order. Accordingly

$$((\phi \otimes \phi) \circ \Delta_C)(a_{i_1} \otimes \cdots \otimes a_{i_p} \otimes b_{j_1} \otimes \cdots \otimes b_{j_q})$$
$$= \sum e(\{a'_1 \otimes \cdots \otimes b'_q\}, \{a''_1 \otimes \cdots \otimes b''_q\})(a' \otimes b') \otimes (a'' \otimes b''),$$

where $a' = a'_1 \otimes \cdots \otimes a'_p$, $a'' = a''_1 \otimes \cdots \otimes a''_p$, $b' = b'_1 \otimes \cdots \otimes b'_q$ and $b'' = b''_1 \otimes \cdots \otimes b''_q$.

Put $I' = \{a'_1 \otimes \cdots \otimes a'_p\}$, $J' = \{b'_1 \otimes \cdots \otimes b'_q\}$ and define I'' and J'' similarly. Also, if $I' = (i'_1, i'_2, \ldots, i'_p)$ and $J' = (j'_1, j'_2, \ldots, j'_q)$ let us use $I'J'$ to denote the sequence $(i'_1, \ldots, i'_p, j'_1, \ldots, j'_q)$. We then have

$$((\phi \otimes \phi) \circ \Delta_C)(a_{i_1} \otimes \cdots \otimes a_{i_p} \otimes b_{j_1} \otimes \cdots \otimes b_{j_q})$$
$$= \sum e(I'J', I''J'')(a' \otimes b') \otimes (a'' \otimes b''). \tag{7.5.14}$$

But, on the other hand,

$$\Delta_A(a_{i_1} \otimes \cdots \otimes a_{i_p}) = \sum e(\{a'\}, \{a''\})(a' \otimes a'') = \sum e(I', I'')(a' \otimes a'')$$

and

$$\Delta_B(b_{j_1} \otimes b_{j_2} \otimes \cdots \otimes b_{j_q})$$
$$= \sum e(\{b'\}, \{b''\})(b' \otimes b'') = \sum e(J', J'')(b' \otimes b'')$$

so therefore

$$(\Delta_A \otimes \Delta_B)((a_{i_1} \otimes \cdots \otimes a_{i_p}) \otimes (b_{j_1} \otimes \cdots \otimes b_{j_q}))$$
$$= \sum e(I', I'')e(J', J'')(a' \otimes a'') \otimes (b' \otimes b'').$$

Now a', a'' are homogeneous elements of A and b', b'' are homogeneous elements of B. Furthermore, if $J' = (j'_1, \ldots, j'_q)$ and $I'' = (i''_1, \ldots, i''_p)$, then the degree of b' is $|J'| = j'_1 + j'_2 + \cdots + j'_q$ and that of a'' is $|I''| = i''_1 + i''_2 + \cdots + i''_p$. Accordingly

$$\Delta_{A \underline{\otimes} B}((a_{i_1} \otimes \cdots \otimes a_{i_p}) \otimes (b_{j_1} \otimes \cdots \otimes b_{j_q}))$$
$$= \sum e(I', I'')e(J', J'')(-1)^{|J'||I''|}(a' \otimes b') \otimes (a'' \otimes b''). \tag{7.5.15}$$

However, it is easy to check that

$$e(I', I'')e(J', J'')(-1)^{|J'||I''|} = e(I'J', I''J'')$$

and so, comparing (7.5.14) with (7.5.15), we see that $(\phi \otimes \phi) \circ \Delta_C$ and $\Delta_{A \underline{\otimes} B} \circ \phi$ produce the same result when applied to $a_{i_1} \otimes \cdots \otimes a_{i_p} \otimes b_{j_1} \otimes \cdots \otimes b_{j_q}$. It follows that $(\phi \otimes \phi) \circ \Delta_C = \Delta_{A \underline{\otimes} B} \circ \phi$ and with this the theorem is proved.

The final result in this section shows that the formation of modified tensor products of coalgebras is a commutative operation. To prepare the way for proving this, first suppose that A and B are graded R-modules with $\{A_n\}_{n\in\mathbb{Z}}$ the grading on A and $\{B_n\}_{n\in\mathbb{Z}}$ the grading on B. We recall that the *twisting isomorphism* (see (3.8.5))

$$T: A \otimes B \xrightarrow{\sim} B \otimes A \tag{7.5.16}$$

is an isomorphism of R-modules in which

$$T(a_m \otimes b_n) = (-1)^{mn} b_n \otimes a_m. \tag{7.5.17}$$

Here, of course, $a_m \in A_m$ and $b_n \in B_n$.

Now assume that A and B are graded coalgebras. In this case we can regard the twisting isomorphism as mapping the coalgebra $A \underline{\otimes} B$ onto the coalgebra $B \underline{\otimes} A$. The next theorem shows that, in this context, T is an isomorphism of coalgebras.

Theorem 6. *Let A and B be graded coalgebras. Then the R-linear bijection $T: A \underline{\otimes} B \to B \underline{\otimes} A$, where T is the twisting isomorphism, is an isomorphism of graded coalgebras.*

Proof. T certainly preserves degrees. Now suppose that $a_m \in A_m$ and $b_n \in B_n$. Then

$$\begin{aligned}\varepsilon_{B \underline{\otimes} A}(T(a_m \otimes b_n)) &= (-1)^{mn} \varepsilon_{B \underline{\otimes} A}(b_n \otimes a_m) \\ &= (-1)^{mn} \varepsilon_B(b_n) \varepsilon_A(a_m) \\ &= \varepsilon_B(b_n) \varepsilon_A(a_m)\end{aligned}$$

because, since A and B are graded coalgebras, $\varepsilon_B(b_n)\varepsilon_A(a_m)$ is zero except perhaps when $m=n=0$. It follows that $\varepsilon_{B \underline{\otimes} A}(T(a_m \otimes b_n)) = \varepsilon_{A \underline{\otimes} B}(a_m \otimes b_n)$. Accordingly $\varepsilon_{B \underline{\otimes} A} \circ T = \varepsilon_{A \underline{\otimes} B}$ and we have shown that T preserves counits.

We turn now to the question of compatibility with comultiplication. We can write $\Delta_A(a_m) = \sum a' \otimes a''$, where a', a'' are homogeneous elements of A the sum of whose degrees is m; likewise we have $\Delta_B(b_n) = \sum b' \otimes b''$, where this time b', b'' are homogeneous elements of B and the sum of their degrees is n. Let us use $\{a'\}, \{a''\}, \{b'\}, \{b''\}$ to denote the degrees of these various elements. Then

$$\Delta_{A \underline{\otimes} B}(a_m \otimes b_n) = \sum (-1)^{\{b'\}\{a''\}} (a' \otimes b') \otimes (a'' \otimes b'')$$

and therefore

$$\begin{aligned}&(T \otimes T)(\Delta_{A \underline{\otimes} B}(a_m \otimes b_n)) \\ &\quad= \sum (-1)^{\{b'\}\{a''\}} (-1)^{\{b'\}\{a'\}} (-1)^{\{b''\}\{a''\}} (b' \otimes a') \otimes (b'' \otimes a'') \\ &\quad= (-1)^{mn} \sum (-1)^{\{a'\}\{b''\}} (b' \otimes a') \otimes (b'' \otimes a'')\end{aligned}$$

$$=(-1)^{mn}\Delta_{B\otimes A}(b_n\otimes a_m)$$
$$=\Delta_{B\otimes A}(T(a_m\otimes b_n)).$$

This shows that $(T\otimes T)\circ\Delta_{A\otimes B}=\Delta_{B\otimes A}\circ T$ and now the proof is complete.

7.6 Commutative and skew-commutative coalgebras

Let A and B be R-modules, and let us use

$$U: A\otimes B \xrightarrow{\sim} B\otimes A \tag{7.6.1}$$

to denote the module-isomorphism in which $U(a\otimes b)=b\otimes a$. If now (A,μ,η) is an R-algebra, then it is a *commutative* algebra if and only if

$$\begin{array}{ccc} A\otimes A & \xrightarrow{U} & A\otimes A \\ & {\scriptstyle\mu}\searrow \quad \swarrow{\scriptstyle\mu} & \\ & A & \end{array} \tag{7.6.2}$$

is a commutative diagram.

Suppose next that (A,μ,η) is a *graded* algebra and let $T: A\otimes A \xrightarrow{\sim} A\otimes A$ be the twisting isomorphism (see (7.5.16)). We shall say that A is a *skew-commutative* algebra provided that the diagram

$$\begin{array}{ccc} A\otimes A & \xrightarrow{T} & A\otimes A \\ & {\scriptstyle\mu}\searrow \quad \swarrow{\scriptstyle\mu} & \\ & A & \end{array} \tag{7.6.3}$$

commutes. This is equivalent to requiring that

$$a_m\alpha_n=(-1)^{mn}\alpha_n a_m \tag{7.6.4}$$

for homogeneous elements a_m, α_n of degrees m and n respectively. Note that this is weaker than the requirement of being *anticommutative*; in fact a graded algebra is anticommutative if and only if (i) it is skew-commutative, and (ii) the square of every homogeneous element of odd degree is zero.

These observations suggest the following definitions for coalgebras. A coalgebra (A,Δ,ε) will be called *commutative* if the diagram

commutes. Should the coalgebra be graded it will be called *skew-commutative* provided that

(7.6.6)

is a commutative diagram. Here, of course, T is the twisting isomorphism.

With this definition we have reached the point where the parallelism between algebras and coalgebras has been developed as far as we need. Further results will be found in the exercises at the end of the chapter.

7.7 Linear forms on a coalgebra

It will now be shown that the linear forms on a coalgebra constitute, in a natural way, an R-algebra. To see how this comes about, suppose that (A, Δ, ε) is a given R-coalgebra and put

$$A^* = \mathrm{Hom}_R(A, R) \tag{7.7.1}$$

so that A^* is the module of linear forms on A. First we note that the counit ε belongs to A^* and therefore there exists an R-linear mapping

$$\eta: R \to A^* \tag{7.7.2}$$

in which $\eta(r) = r\varepsilon$. This mapping will provide the structural homomorphism of our algebra.

Next let $\mu_R: R \otimes R \to R$ be the multiplication mapping of R (considered as an algebra) and let f, g belong to A^*. Then $\mu_R \circ (f \otimes g) \circ \Delta$ also belongs to A^* and the mapping $A^* \times A^* \to A^*$ in which (f, g) becomes $\mu_R \circ (f \otimes g) \circ \Delta$ is bilinear. It follows that there is a homomorphism

$$\mu: A^* \otimes A^* \to A^* \tag{7.7.3}$$

of R-modules which is such that

$$\mu(f \otimes g) = \mu_R \circ (f \otimes g) \circ \Delta. \tag{7.7.4}$$

Theorem 7. *Let (A, Δ, ε) be an R-coalgebra. Then (with the above notation) the triplet (A^*, μ, η) is an R-algebra.*

Proof. Let f, g, h belong to A^* and consider the mappings

$$A \longrightarrow A \otimes A \otimes A \xrightarrow{f \otimes g \otimes h} R \otimes R \otimes R \longrightarrow R, \tag{7.7.5}$$

where $A \to A \otimes A \otimes A$ is $(\Delta \otimes A) \circ \Delta = (A \otimes \Delta) \circ \Delta$ and $R \otimes R \otimes R \to R$ takes $r_1 \otimes r_2 \otimes r_3$ into $r_1 r_2 r_3$. The total mapping (7.7.5) can be factored into

$$A \xrightarrow{\Delta} A \otimes A \xrightarrow{\mu(f \otimes g) \otimes A} R \otimes A \xrightarrow{R \otimes h} R \otimes R \xrightarrow{\mu_R} R$$

and hence into

$$A \xrightarrow{\Delta} A \otimes A \xrightarrow{\mu(f \otimes g) \otimes h} R \otimes R \xrightarrow{\mu_R} R.$$

But this is just $\mu(\mu(f \otimes g) \otimes h)$. In a similar manner we can verify that the complete mapping (7.7.5) is also the same as $\mu(f \otimes \mu(g \otimes h))$. Consequently, if we use μ to define multiplication on A^*, then multiplication will be associative.

Next, by (7.7.4),

$$\mu(\varepsilon \otimes f) = \mu_R \circ (\varepsilon \otimes f) \circ \Delta = \mu_R \circ (R \otimes f) \circ (\varepsilon \otimes A) \circ \Delta$$

and, because (A, Δ, ε) is a coalgebra, $(\varepsilon \otimes A) \circ \Delta$ is the canonical isomorphism of A onto $R \otimes A$. It follows that $\mu(\varepsilon \otimes f): A \to R$ takes a into $f(a)$ and therefore $\mu(\varepsilon \otimes f) = f$. A similar argument shows that $\mu(f \otimes \varepsilon) = f$ as well. It is now apparent that A^* is an associative R-algebra, with multiplication μ and identity element ε. Furthermore $\eta: R \to A^*$ is its structural homomorphism.

Corollary. *If* (A, Δ, ε) *is a commutative coalgebra, then* (A^*, μ, η) *is a commutative algebra.*

Proof. Let f, g belong to A^*. If now $a, a' \in A$, then the total mapping

$$A \otimes A \xrightarrow{U} A \otimes A \xrightarrow{f \otimes g} R \otimes R \xrightarrow{\mu_R} R$$

takes $a \otimes a'$ into $f(a')g(a)$. Accordingly

$$\mu_R \circ (f \otimes g) \circ U = \mu_R \circ (g \otimes f).$$

But $\Delta = U \circ \Delta$ because (A, Δ, ε) is commutative. Consequently

$$\mu_R \circ (f \otimes g) \circ \Delta = \mu_R \circ (g \otimes f) \circ \Delta,$$

i.e. $\mu(f \otimes g) = \mu(g \otimes f)$ which is what we were seeking to prove.

We next turn our attention to the case where (A, Δ, ε) is endowed with a coalgebra grading $\{A_n\}_{n \in \mathbb{Z}}$. For a given integer k, the linear forms on A that vanish on $\sum_{n \neq k} A_n$ form a submodule of A^*; moreover this submodule is isomorphic to $\mathrm{Hom}_R(A_k, R) = A_k^*$ under an isomorphism which maps the relevant linear forms on A into their restrictions on A_k. We may therefore regard A_k^* as an R-submodule of A^*. On this understanding put

$$\mathfrak{A} = \sum_{k \in \mathbb{Z}} A_k^* \tag{7.7.6}$$

so that $\mathfrak{A}$ is a submodule of A^*. Note that the sum (7.7.6) is *direct*. Usually A^* and $\mathfrak{A}$ are different, but *in the special case where only finitely many* A_k *are non-zero we always have* $A^* = \mathfrak{A}$.

It is clear that $\varepsilon \in A_0^*$ because ε preserves degrees. Assume now that $f \in A_p^*$ and $g \in A_q^*$. Then $f \otimes g$, which of course maps $A \otimes A$ into $R \otimes R$, vanishes on $A_s \otimes A_t$ unless both $s = p$ and $t = q$. But $\Delta: A \to A \otimes A$ preserves degrees and from this it follows that $\mu(f \otimes g) \in A_{p+q}^*$. We now see that $\mathfrak{A}$ can be regarded as a graded subalgebra of the algebra (A, μ, η) described in Theorem 7. More precisely we have proved

Theorem 8. *Let (A, Δ, ε) be a graded R-coalgebra with $\{A_n\}_{n\in\mathbb{Z}}$ as its grading. Then (with the above notation) $\mathfrak{A}$ is an R-algebra and it has $\{A_n^*\}_{n\in\mathbb{Z}}$ as an algebra-grading. (Here A_n^* is the dual of A_n and it is to be considered as a submodule of the dual, A^*, of A.)*

Corollary. *Let the graded R-coalgebra (A, Δ, ε) be skew-commutative. Then $\mathfrak{A}$ is a skew-commutative algebra.*

Proof. Let $f \in A_p^*$ and $g \in A_q^*$. It is easily checked that if $T: A \otimes A \xrightarrow{\sim} A \otimes A$ denotes the twisting isomorphism, then

$$\mu_R \circ (f \otimes g) \circ T = (-1)^{pq} \mu_R \circ (g \otimes f).$$

But $T \circ \Delta = \Delta$ because (A, Δ, ε) is skew-commutative, and now it follows that $\mu(f \otimes g) = (-1)^{pq} \mu(g \otimes f)$.

7.8 Hopf algebras

Let A be an R-module and suppose that we are given R-linear mappings as follows:

$$\mu: A \otimes A \rightarrow A, \tag{7.8.1}$$

$$\eta: R \rightarrow A, \tag{7.8.2}$$

$$\Delta: A \rightarrow A \otimes A, \tag{7.8.3}$$

$$\varepsilon: A \rightarrow R. \tag{7.8.4}$$

Then A and these various homomorphisms are said to constitute a *Hopf algebra* provided that the four conditions

(i) *(A, μ, η) is an R-algebra;*
(ii) *(A, Δ, ε) is an R-coalgebra;*
(iii) *$\Delta: A \rightarrow A \otimes A$ and $\varepsilon: A \rightarrow R$ are homomorphisms of algebras;*
(iv) *$\mu: A \otimes A \rightarrow A$ and $\eta: R \rightarrow A$ are homomorphisms of coalgebras;*

are all satisfied. When this is the case we shall speak of the Hopf algebra $(\eta, \mu, A, \Delta, \varepsilon)$.

The conditions (i)–(iv) are not independent; indeed the precise connections between them are explained in the following lemma. Note that if (i) and (ii) are both satisfied, then $A \otimes A$ is both an algebra and a coalgebra.

Lemma 5. *Suppose that (A, μ, η) is an algebra and that (A, Δ, ε) is a coalgebra. Then*

(*a*) *$\mu: A \otimes A \rightarrow$ is compatible with comultiplication if and only if $\Delta: A \rightarrow A \otimes A$ is compatible with multiplication;*

(*b*) *$\mu: A \otimes A \to A$ preserves counits if and only if $\varepsilon: A \to R$ is compatible with multiplication;*

(*c*) *$\eta: R \to A$ is compatible with comultiplication if and only if $\Delta: A \to A \otimes A$ preserves identities;*

(*d*) *$\eta: R \to A$ preserves counits if and only if $\varepsilon: A \to R$ preserves identities.*

Proof. (*a*) The mapping $\mu: A \otimes A \to A$ is compatible with comultiplication if and only if $\Delta \circ \mu = (\mu \otimes \mu) \circ \Delta_{A \otimes A}$. But

$$\Delta_{A \otimes A} = (A \otimes U \otimes A) \circ (\Delta \otimes \Delta),$$

where $U: A \otimes A \xrightarrow{\sim} A \otimes A$ is the isomorphism in which $U(a \otimes a') = a' \otimes a$. Consequently our condition becomes

$$\Delta \circ \mu = (\mu \otimes \mu) \circ (A \otimes U \otimes A) \circ (\Delta \otimes \Delta). \tag{7.8.5}$$

On the other hand $\Delta: A \to A \otimes A$ is compatible with multiplication precisely when $\Delta \circ \mu = \mu_{A \otimes A} \circ (\Delta \otimes \Delta)$ and this is equivalent to (7.8.5) because $\mu_{A \otimes A} = (\mu \otimes \mu) \circ (A \otimes U \otimes A)$.

(*b*) For $\mu: A \otimes A \to A$ to preserve counits it is necessary and sufficient that $\varepsilon \circ \mu = \varepsilon_{A \otimes A}$, that is to say we require $\varepsilon \circ \mu = \mu_R \circ (\varepsilon \otimes \varepsilon)$. However, this is precisely the condition for $\varepsilon: A \to R$ to be compatible with multiplication.

(*c*) The structural homomorphism $\eta: R \to A$ is compatible with comultiplication if and only if $(\eta \otimes \eta) \circ \Delta_R = \Delta \circ \eta$, and this occurs when and only when $\Delta(1_A) = 1_A \otimes 1_A$.

(*d*) For $\eta: R \to A$ to preserve counits we require that $\varepsilon_R = \varepsilon \circ \eta$ and this occurs precisely when $\varepsilon(1_A) = 1_R$.

We combine some of these observations in the following

Corollary. *Suppose that (A, μ, η) is an algebra and that (A, Δ, ε) is a coalgebra. Then $\mu: A \otimes A \to A$ and $\eta: R \to A$ are homomorphisms of coalgebras if and only if $\Delta: A \to A \otimes A$ and $\varepsilon: A \to R$ are homomorphisms of algebras.*

Now suppose that $(\eta_A, \mu_A, A, \Delta_A, \varepsilon_A)$ and $(\eta_B, \mu_B, B, \Delta_B, \varepsilon_B)$ are Hopf algebras. A mapping $A \to B$ is called a *homomorphism of Hopf algebras* provided it is homomorphism both of algebras and of coalgebras. A bijective homomorphism of Hopf algebras is called an *isomorphism* of Hopf algebras. Evidently if $f: A \to B$ is such an isomorphism, then $f^{-1}: B \to A$ is also an isomorphism of Hopf algebras. Again, if $g: A \to B$ and $h: B \to C$ are homomorphisms of Hopf algebras, then so is their product $h \circ g$.

Another concept which we shall need is that of a commutative Hopf algebra. Quite simply, a Hopf algebra is said to be *commutative* if it is commutative both as an algebra and as a coalgebra.

Next assume that $(\eta, \mu, A, \Delta, \varepsilon)$ is a Hopf algebra and let $\{A_n\}_{n \in \mathbb{Z}}$ be a family of R-submodules of A which grades A as an R-module. We then say that $\{A_n\}_{n\in\mathbb{Z}}$ *grades the Hopf algebra* provided that it grades the algebra (A, μ, η) and the coalgebra (A, Δ, ε). This will be the case if $\mu, \eta, \Delta, \varepsilon$ all preserve the degrees of homogeneous elements. (Naturally, it is understood that $A \otimes A$ has the total grading and that R is graded trivially.) Finally, if A and B are graded Hopf algebras and $f: A \to B$ is a degree-preserving homomorphism (of Hopf algebras), then we say that f is a *homomorphism of graded Hopf algebras.*

In the presence of a grading there is an important way in which our definition of a Hopf algebra can be modified. In order to explain how this comes about we shall make a completely fresh start.

Assume then that A is an R-module and that R-linear mappings $\mu: A \otimes A \to A$, $\eta: R \to A$, $\Delta: A \to A \otimes A$, and $\varepsilon: A \to R$ are given. Suppose also that A is graded, as an R-module, by $\{A_n\}_{n\in\mathbb{Z}}$. Should it happen that this makes (A, μ, η) a graded algebra and (A, Δ, ε) a graded coalgebra, then the twisted product $A \underline{\otimes} A$ will exist both as a graded algebra and as a graded coalgebra. Furthermore, in these circumstances we can regard Δ as a mapping of A into $A \underline{\otimes} A$ and μ as a mapping of $A \underline{\otimes} A$ into A. With this in mind, we say that the complete system forms a *modified Hopf algebra* or a *twisted Hopf algebra* provided that $\Delta: A \to A \underline{\otimes} A$ and $\varepsilon: A \to R$ are homomorphisms of (graded) R-algebras and $\mu: A \underline{\otimes} A \to A$ and $\eta: R \to A$ are homomorphisms of (graded) R-coalgebras. Thus to modify the definition of a Hopf algebra we introduce a grading and then replace $A \otimes A$ by $A \underline{\otimes} A$.

For the new situation the counterpart of Lemma 5 is

Lemma 6. *Let (A, μ, η) be an algebra, let (A, Δ, ε) be a coalgebra, and let both the algebra and the coalgebra be graded by $\{A_n\}_{n\in\mathbb{Z}}$. Then*

(*a*) *$\mu: A \underline{\otimes} A \to A$ is compatible with comultiplication if and only if $\Delta: A \to A \underline{\otimes} A$ is compatible with multiplication;*

(*b*) *$\mu: A \underline{\otimes} A \to A$ preserves counits if and only if $\varepsilon: A \to R$ is compatible with multiplication;*

(*c*) *$\eta: R \to A$ is compatible with comultiplication if and only if $\Delta: A \to A \underline{\otimes} A$ preserves identities;*

(*d*) *$\eta: R \to A$ preserves counits if and only if $\varepsilon: A \to R$ preserves identities.*

Proof. Because of the close similarity of this result to Lemma 5 we shall only prove (*a*). For this we observe that $\mu: A \underline{\otimes} A \to A$ is compatible with comultiplication if and only if $\Delta \circ \mu = (\mu \otimes \mu) \circ \Delta_{A \underline{\otimes} A}$. Now

$$\Delta_{A \underline{\otimes} A} = (A \otimes T \otimes A) \circ (\Delta \otimes \Delta),$$

where $T: A \otimes A \xrightarrow{\sim} A \otimes A$ is the twisting isomorphism, so that what we require is

$$\Delta \circ \mu = (\mu \otimes \mu) \circ (A \otimes T \otimes A) \circ (\Delta \otimes \Delta). \tag{7.8.6}$$

On the other hand, for $\Delta: A \to A \underline{\otimes} A$ to be compatible with multiplication it is necessary and sufficient that $\Delta \circ \mu = \mu_{A \underline{\otimes} A} \circ (\Delta \otimes \Delta)$ and this is equivalent to (7.8.6) because $\mu_{A \underline{\otimes} A} = (\mu \otimes \mu) \circ (A \otimes T \otimes A)$.

Corollary. *Suppose that (A, μ, η) is an algebra and (A, Δ, ε) is a coalgebra, and let both be graded by $\{A_n\}_{n \in \mathbb{Z}}$. Then $\mu: A \underline{\otimes} A \to A$ and $\eta: R \to A$ are homomorphisms of (graded) coalgebras if and only if $\Delta: A \to A \underline{\otimes} A$ and $\varepsilon: A \to R$ are homomorphisms of (graded) algebras.*

For modified Hopf algebras, the definitions of homomorphism and isomorphism are the same as for ordinary (i.e. unmodified) Hopf algebras, except that such mappings are now understood to preserve degrees. A modified Hopf algebra is termed *skew-commutative* if its algebra and coalgebra components are both skew-commutative.

7.9 Tensor products of Hopf algebras

For $1 \leq i \leq p$ let $(\eta_i, \mu_i, A_i, \Delta_i, \varepsilon_i)$ be a Hopf algebra (over the ring R) and put

$$A = A_1 \otimes A_2 \otimes \cdots \otimes A_p. \tag{7.9.1}$$

Then A has a natural structure both as an algebra and as a coalgebra. Indeed, by (7.1.5) and (7.1.6),

$$\mu_A = (\mu_1 \otimes \mu_2 \otimes \cdots \otimes \mu_p) \circ \Lambda_p, \tag{7.9.2}$$

$$\eta_A = (\eta_1 \otimes \eta_2 \otimes \cdots \otimes \eta_p) \circ \Delta_R^{(p)}, \tag{7.9.3}$$

and, by (7.4.4) and (7.4.7),

$$\Delta_A = V_p \circ (\Delta_1 \otimes \Delta_2 \otimes \cdots \otimes \Delta_p), \tag{7.9.4}$$

$$\varepsilon_A = \mu_R^{(p)} \circ (\varepsilon_1 \otimes \varepsilon_2 \otimes \cdots \otimes \varepsilon_p). \tag{7.9.5}$$

(The notation is explained in Sections (7.1) and (7.4).) Now, because A_i is a Hopf algebra, $\Delta_1 \otimes \Delta_2 \otimes \cdots \otimes \Delta_p$ is an algebra-homomorphism of $A_1 \otimes A_2 \otimes \cdots \otimes A_p$ into $(A_1 \otimes A_1) \otimes (A_2 \otimes A_2) \otimes \cdots \otimes (A_p \otimes A_p)$, and it follows, from Theorems 2 and 3 of Chapter 3, that V_p is an algebra-isomorphism of

$$(A_1 \otimes A_1) \otimes (A_2 \otimes A_2) \otimes \cdots \otimes (A_p \otimes A_p)$$

onto $(A_1 \otimes A_2 \otimes \cdots \otimes A_p) \otimes (A_1 \otimes A_2 \otimes \cdots \otimes A_p)$. Accordingly $\Delta_A: A \to A \otimes A$ is a homomorphism of R-algebras. Next $\mu_R^{(p)}$ (see (7.4.5)) is a homomorphism of R-algebras and, of course, $\varepsilon_1 \otimes \varepsilon_2 \otimes \cdots \otimes \varepsilon_p$ is an algebra-homomorphism of $A_1 \otimes A_2 \otimes \cdots \otimes A_p$ into $R \otimes R \otimes \cdots \otimes R$. Thus $\varepsilon_A: A \to R$ is a homomorphism of algebras as well.

Theorem 9. *Let $A_1, A_2, \ldots, A_p$ be Hopf algebras over R. Put $A = A_1 \otimes A_2 \otimes \cdots \otimes A_p$ and define $\mu_A, \eta_A, \Delta_A, \varepsilon_A$ as above. Then $(\eta_A, \mu_A, A, \Delta_A, \varepsilon_A)$ is a Hopf algebra.*

Proof. The theorem follows at once from the fact that $\Delta_A: A \to A \otimes A$ and $\varepsilon_A: A \to R$ are algebra-homomorphisms by applying the corollary to Lemma 5.

Definition. *The Hopf algebra $(\eta_A, \mu_A, A, \Delta_A, \varepsilon_A)$ of the theorem is called the 'tensor product' of the Hopf algebras $A_1, A_2, \ldots, A_p$ and it is denoted by $A_1 \otimes A_2 \otimes \cdots \otimes A_p$.*

The corresponding result for modified Hopf algebras presents minor complications. The following lemma will help us to deal with these.

Lemma 7. *Let A and B be modified Hopf algebras. Then $A \underline{\otimes} B$, which is certainly both a graded algebra and a graded coalgebra, is in fact a modified Hopf algebra.*

Proof. The comultiplication $\Delta_{A \underline{\otimes} B}$, of the coalgebra $A \underline{\otimes} B$, can be regarded as a mapping of $A \underline{\otimes} B$ into $(A \underline{\otimes} B) \underline{\otimes} (A \underline{\otimes} B)$. Furthermore, it is given by

$$\Delta_{A \underline{\otimes} B} = (A \otimes T \otimes B) \circ (\Delta_A \otimes \Delta_B), \tag{7.9.6}$$

where $T: A \otimes B \xrightarrow{\sim} B \otimes A$ is the twisting isomorphism. Next, (7.9.6) can be obtained by piecing together the mappings

$$A \underline{\otimes} B \xrightarrow{\Delta_A \otimes \Delta_B} (A \underline{\otimes} A) \underline{\otimes} (B \underline{\otimes} B) \approx A \underline{\otimes} (A \underline{\otimes} B) \underline{\otimes} B$$

$$\xrightarrow{A \otimes T \otimes B} A \underline{\otimes} (B \underline{\otimes} A) \underline{\otimes} B \approx (A \underline{\otimes} B) \underline{\otimes} (A \underline{\otimes} B), \tag{7.9.7}$$

where the unlabelled isomorphisms are the obvious module-isomorphisms based on Theorem 1 of Chapter 2. However, $\Delta_A: A \to A \underline{\otimes} A$ and $\Delta_B: B \to B \underline{\otimes} B$ are algebra-homomorphisms by hypothesis and $T: A \underline{\otimes} B \xrightarrow{\sim} B \underline{\otimes} A$ is an algebra-isomorphism by Exercise 5 of Chapter 3. It follows that all the mappings in (7.9.7) are homomorphisms of algebras and therefore

$$\Delta_{A \underline{\otimes} B}: A \underline{\otimes} B \to (A \underline{\otimes} B) \underline{\otimes} (A \underline{\otimes} B)$$

is a homomorphism of algebras as well.

We turn now to $\varepsilon_{A \underline{\otimes} B}$. This is the result of combining the mappings

$$A \underline{\otimes} B \xrightarrow{\varepsilon_A \otimes \varepsilon_B} R \underline{\otimes} R \xrightarrow{\mu_R^{(2)}} R.$$

Since $R \underline{\otimes} R$ and $R \otimes R$ are identical as algebras, we see that $\varepsilon_{A \underline{\otimes} B}: A \underline{\otimes} B \to R$ is also an algebra-homomorphism. Finally, to complete the proof we have only to apply the corollary to Lemma 6.

Theorem 10. *Let $A^{(1)}, A^{(2)}, \ldots, A^{(p)}$ be modified Hopf algebras. Then the algebra and coalgebra structures on $A^{(1)} \underline{\otimes} A^{(2)} \underline{\otimes} \cdots \underline{\otimes} A^{(p)}$ interact in such a way as to make it a modified Hopf algebra.*

Proof. We use induction on p. When $p=1$ the statement is tautologous so we shall assume that $p>1$ and that the result in question has been proved in the case of $p-1$ modified Hopf algebras. This secures that $A^{(1)} \underline{\otimes} A^{(2)} \underline{\otimes} \cdots \underline{\otimes} A^{(p-1)}$ is a modified Hopf algebra and therefore, by Lemma 7, the same is true of

$$(A^{(1)} \underline{\otimes} A^{(2)} \underline{\otimes} \cdots \underline{\otimes} A^{(p-1)}) \underline{\otimes} A^{(p)}.$$

Next, Theorem 1 of Chapter 2 provides a degree-preserving *module-isomorphism*

$$A^{(1)} \underline{\otimes} \cdots \underline{\otimes} A^{(p-1)} \underline{\otimes} A^{(p)}$$
$$\approx (A^{(1)} \underline{\otimes} \cdots \underline{\otimes} A^{(p-1)}) \underline{\otimes} A^{(p)} \quad (7.9.8)$$

However, by Theorem 6 of Chapter 3, this is an isomorphism of graded algebras, and (this time by Theorem 5 of the present chapter) it is an isomorphism of graded coalgebras. Since the right-hand side of (7.9.8) is a modified Hopf algebra, we may conclude that the same holds for the left-hand side. The inductive step is now complete and the theorem follows.

7.10 $E(M)$ as a (modified) Hopf algebra

Let M be an R-module. We already know that $E(M)$ is an anticommutative, graded R-algebra and that $S(M)$ is a commutative, graded R-algebra. But this is far from being the whole story. It will be shown, in this section, that the structure on $E(M)$ can be extended so that $E(M)$ becomes a modified, skew-commutative, Hopf algebra; and later we shall establish that $S(M)$ is a commutative, graded, Hopf algebra. Our reason for dealing with $E(M)$ first is because the details in this case are a little more complicated. When we come to discuss $S(M)$ in a similar manner, the reader will find that there is no difficulty in adapting the treatment provided for exterior algebras.

Suppose then that we are given an R-module M. Let us form its exterior algebra $E(M)$ and let us identify $E_0(M)$ with R. The multiplication mapping $\mu: E(M) \otimes E(M) \to E(M)$ of the algebra satisfies $\mu(x \otimes y) = x \wedge y$ and the unit mapping $\eta: R \to E(M)$ is simply the identity mapping of R onto $E_0(M)$.

Now let

$$\varepsilon: E(M) \to R \quad (7.10.1)$$

be the projection of $E(M)$ onto $E_0(M)$. This mapping will turn out to be the counit of $E(M)$ when we come to consider the latter as a coalgebra. Note that ε *is a homomorphism of graded R-algebras.*

The definition of comultiplication requires some preparation. The modified tensor product $E(M)\underline{\otimes}E(M)$ of $E(M)$ with itself is an anti-commutative R-algebra (see Chapter 3, Theorem 7). Consequently the mapping $\phi: M\to E(M)\underline{\otimes}E(M)$ in which $\phi(m)=m\otimes 1+1\otimes m$ is not only R-linear, but also satisfies $(\phi(m))^2=0$ for all m in M. Consequently ϕ can be extended to a *homomorphism*

$$\Delta: E(M)\to E(M)\underline{\otimes}E(M) \tag{7.10.2}$$

of R-algebras which is degree-preserving and which is such that

$$\Delta(m)=m\otimes 1+1\otimes m \tag{7.10.3}$$

when $m\in M$. The homomorphism Δ is called the *diagonalization mapping* of $E(M)$. Of course, since $E(M)\underline{\otimes}E(M)$ and $E(M)\otimes E(M)$ coincide as modules (but not as algebras), Δ can also be considered as an R-linear mapping of $E(M)$ into $E(M)\otimes E(M)$.

Lemma 8. *With the above notation the triple* $(E(M),\Delta,\varepsilon)$ *is a coalgebra. This is graded, as a coalgebra, by the exterior powers* $\{E_n(M)\}_{n\in\mathbb{Z}}$ *of the module M. As a graded coalgebra* $(E(M),\Delta,\varepsilon)$ *is skew-commutative.*

Proof. To see that the triple constitutes a coalgebra it suffices to verify that the diagrams

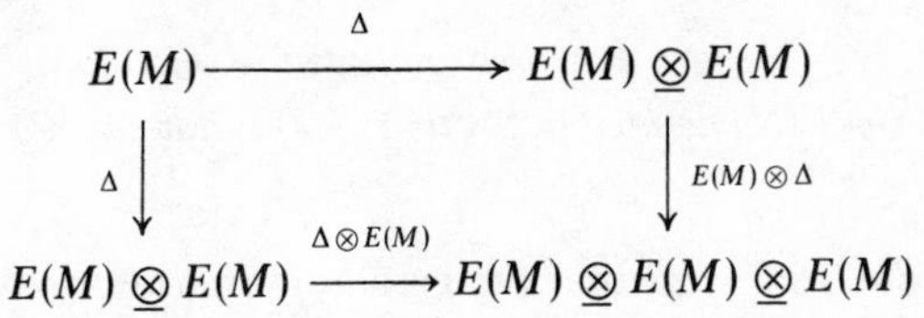

$$\begin{array}{ccc} E(M) & \xrightarrow{\quad\Delta\quad} & E(M)\underline{\otimes}E(M) \\ \Big\downarrow{\scriptstyle\Delta} & & \Big\downarrow{\scriptstyle E(M)\otimes\Delta} \\ E(M)\underline{\otimes}E(M) & \xrightarrow{\Delta\otimes E(M)} & E(M)\underline{\otimes}E(M)\underline{\otimes}E(M) \end{array} \tag{7.10.4}$$

and

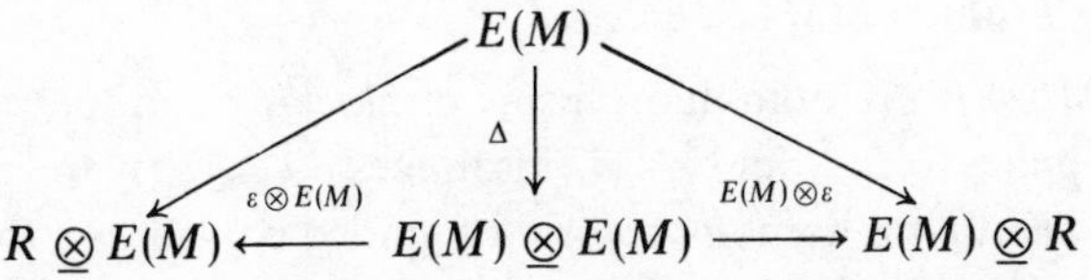

$$\begin{array}{ccccc} & & E(M) & & \\ & \swarrow & \Big\downarrow{\scriptstyle\Delta} & \searrow & \\ R\underline{\otimes}E(M) & \xleftarrow{\varepsilon\otimes E(M)} & E(M)\underline{\otimes}E(M) & \xrightarrow{E(M)\otimes\varepsilon} & E(M)\underline{\otimes}R \end{array} \tag{7.10.5}$$

are commutative. However, all the mappings in (7.10.4) and (7.10.5) are *algebra-homomorphisms* and we know that the R-algebra $E(M)$ is generated by $E_1(M)=M$. Consequently, when verifying that the diagrams are commutative, we need only examine what happens to an element of M. This reduces the verifications to trivialities.

Next, the exterior powers of M grade $E(M)$ as a module and it is obvious that, with respect to this grading, Δ and ε preserve degrees. Accordingly $(E(M),\Delta,\varepsilon)$ is a graded coalgebra.

Finally, to complete the proof we have only to verify that the diagram

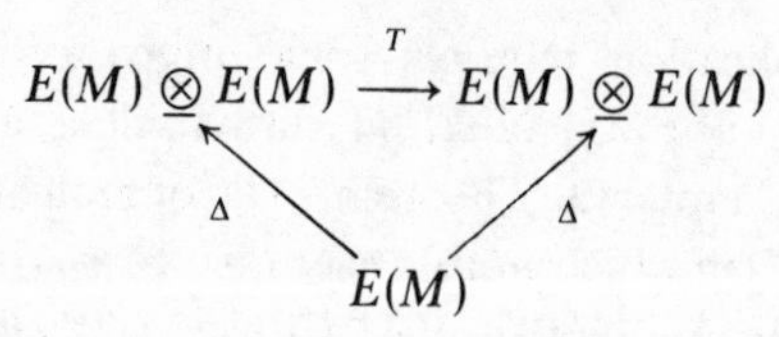

where T is the twisting isomorphism, commutes. But, once again, all the mappings are homomorphisms of algebras so that we need only show that $T(\Delta(m))=\Delta(m)$ for each element m in M. This, however, is clear.

Theorem 11. *Let M be an R-module and let the notation be as above. Then $(\eta, \mu, E(M), \Delta, \varepsilon)$ is a modified Hopf algebra whose grading consists of the exterior powers of M. As a modified Hopf algebra, $E(M)$ is skew-commutative.*

Proof. We know from Chapter 5 that $(E(M), \mu, \eta)$ is a graded algebra and we have just proved (Lemma 8) that $(E(M), \Delta, \varepsilon)$ is a graded coalgebra. Furthermore, we noted earlier that $\Delta: E(M) \to E(M) \underline{\otimes} E(M)$ and $\varepsilon: E(M) \to R$ are homomorphisms of algebras. It therefore follows, from the corollary to Lemma 6, that $(\eta, \mu, E(M), \Delta, \varepsilon)$ is a modified Hopf algebra. This is skew-commutative because $(E(M), \mu, \eta)$ is an anticommutative algebra and, by Lemma 8, $(E(M), \Delta, \varepsilon)$ is a skew-commutative coalgebra. Thus the theorem is proved.

Theorem 12. *Let $f: E \to N$ be a homomorphism of R-modules. Then the mapping $E(f): E(M) \to E(N)$ defined in Section (5.2) is a homomorphism of modified Hopf algebras.*

Proof. We know, from Chapter 5, that $E(f)$ is a homomorphism of graded algebras. We also know that

$$\Delta_{E(M)}: E(M) \to E(M) \underline{\otimes} E(M)$$

and $\varepsilon_{E(M)}: E(M) \to R$ are homomorphisms of *algebras* and that a similar observation applies to $\Delta_{E(N)}$ and $\varepsilon_{E(N)}$. This makes it very easy to verify that $E(f): E(M) \to E(N)$ is a homomorphism of graded coalgebras. But this is all that is needed to complete the proof of the theorem. The details are left to the reader.

7.11 The Grassmann algebra of a module

Let M be an R-module. Then, by the results of the last section, $E(M)$ is a modified Hopf algebra, and so *inter alia* it is a graded coalgebra. We recall that its comultiplication mapping Δ satisfies

$$\Delta(m) = m \otimes 1 + 1 \otimes m$$

for all $m \in M$, and that the counit mapping ε is the projection of $E(M)$ onto

$E_0(M)$. Furthermore, $\Delta: E(M) \to E(M) \underline{\otimes} E(M)$ and $\varepsilon: E(M) \to R$ are homomorphisms of R-algebras.

By Theorem 8 we can now obtain a graded R-algebra by considering the linear forms on $E(M)$. It is this algebra that we are about to investigate.

Suppose that $p \geq 1$ is an integer. In what follows $I = (i_1, i_2, \ldots, i_s)$ will denote a sequence of s $(0 \leq s \leq p)$ integers satisfying $1 \leq i_1 < i_2 < \cdots < i_s \leq p$, and by I' we shall understand the sequence obtained from $(1, 2, \ldots, p)$ by deleting the terms that belong to I. Furthermore, if $I' = (i'_1, i'_2, \ldots)$, then $\operatorname{sgn}(I, I')$ will denote the sign of $(i_1, \ldots, i_s, i'_1, i'_2, \ldots)$ considered as a permutation of $(1, 2, \ldots, p)$.

Now let $m_1, m_2, \ldots, m_p$ belong to M. Put $m_I = m_{i_1} \wedge m_{i_2} \wedge \cdots \wedge m_{i_s}$ and define $m_{I'}$ similarly, it being understood that the natural conventions apply when $s = 0$ and $s = p$.

Lemma 9. *With the above notation*

$$\Delta(m_1 \wedge m_2 \wedge \cdots \wedge m_p) = \sum_I \operatorname{sgn}(I, I') m_I \otimes m_{I'}.$$

Proof. Since $\Delta: E(M) \to E(M) \underline{\otimes} E(M)$ is a homomorphism of algebras, $\Delta(m_1 \wedge m_2 \wedge \cdots \wedge m_p)$ is the product, in $E(M) \underline{\otimes} E(M)$, of $\Delta(m_1), \Delta(m_2), \ldots, \Delta(m_p)$, that is to say it is the product of $(m_1 \otimes 1 + 1 \otimes m_1)$, $(m_2 \otimes 1 + 1 \otimes m_2)$ and so on. Accordingly $\Delta(m_1 \wedge m_2 \wedge \cdots \wedge m_p)$ is the sum, taken over all sequences $I = (i_1, i_2, \ldots, i_s)$, of the product in $E(M) \underline{\otimes} E(M)$ of the elements

$$1 \otimes m_1, \ldots, m_{i_1} \otimes 1, \ldots, m_{i_s} \otimes 1, \ldots, 1 \otimes m_p.$$

But $E(M) \underline{\otimes} E(M)$ is an anticommutative algebra and here we are concerned with homogeneous elements of degree one. Consequently the product in question is equal to $\operatorname{sgn}(I, I')$ times the product of

$$m_{i_1} \otimes 1, m_{i_2} \otimes 1, \ldots, 1 \otimes m_{i'_1}, 1 \otimes m_{i'_2}, \ldots$$

and therefore it has the value $\operatorname{sgn}(I, I') m_I \otimes m_{I'}$. The lemma follows.

We now put

$$E(M)^* = \operatorname{Hom}_R(E(M), R) \tag{7.11.1}$$

so that $E(M)^*$ consists of all the linear forms on $E(M)$; we also put

$$E_n(M)^* = \operatorname{Hom}_R(E_n(M), R) \tag{7.11.2}$$

and identify $E_n(M)^*$ with the module formed by the linear forms on $E(M)$ that vanish on $E_k(M)$ for all $k \neq n$.

By Theorem 7, $E(M)^*$ is an R-algebra whose identity element is the linear form $\varepsilon: E(M) \to R$ which projects $E(M)$ onto $E_0(M)$. For f and g in $E(M)^*$ we shall use $f \wedge g$ to denote their product in this algebra. Naturally $f \wedge g$ is also a linear form on $E(M)$ and, by (7.7.4),

$$f \wedge g = \mu_R \circ (f \otimes g) \circ \Delta. \tag{7.11.3}$$

(We recall that Δ is the comultiplication mapping of $E(M)$ and that μ_R is the multiplication mapping of R.)

Lemma 10. *Let $m_1, m_2, \ldots, m_p$ $(p \geq 0)$ belong to M and let f, g belong to $E(M)^*$. Then (with the same notation as in Lemma 9)*

$$(f \wedge g)(m_1 \wedge m_2 \wedge \cdots \wedge m_p) = \sum_I \operatorname{sgn}(I, I') f(m_I) g(m_{I'}).$$

Remark. When $p=0$ this is to be interpreted as asserting that $(f \wedge g)(1) = f(1)g(1)$.

Proof. The desired result follows by combining (7.11.3) with Lemma 9.

Lemma 11. *Let $f_1, f_2, \ldots, f_p$ $(p \geq 1)$ belong to $M^* = E_1(M)^*$ and let $f_1 \wedge f_2 \wedge \cdots \wedge f_p$ be their product in the algebra $E(M)^*$. Furthermore, let $m_1, m_2, \ldots, m_p$ belong to M. Then*

$$(f_1 \wedge f_2 \wedge \cdots \wedge f_p)(m_1 \wedge m_2 \wedge \cdots \wedge m_p) = \begin{vmatrix} f_1(m_1) & f_1(m_2) & \cdots & f_1(m_p) \\ f_2(m_1) & f_2(m_2) & \cdots & f_2(m_p) \\ \vdots & \vdots & \vdots\vdots\vdots & \vdots \\ f_p(m_1) & f_p(m_2) & \cdots & f_p(m_p) \end{vmatrix}. \tag{7.11.4}$$

Proof. We use induction on p. The assertion is obviously true when $p=1$ so we shall assume that $p>1$ and that the lemma has been established for all smaller values of the inductive variable. Put $g = f_2 \wedge \cdots \wedge f_p$. Since $f_1(m_{\nu_1} \wedge m_{\nu_2} \wedge \cdots \wedge m_{\nu_h}) = 0$ except when $h=1$, it follows, by Lemma 10, that

$$\begin{aligned} &(f_1 \wedge f_2 \wedge \cdots \wedge f_p)(m_1 \wedge m_2 \wedge \cdots \wedge m_p) \\ &= \sum_{i=1}^{p} (-1)^{i+1} f_1(m_i) g(m_1 \wedge \cdots \wedge \hat{m}_i \wedge \cdots \wedge m_p) \\ &= \begin{vmatrix} f_1(m_1) & f_1(m_2) & \cdots & f_1(m_p) \\ f_2(m_1) & f_2(m_2) & \cdots & f_2(m_p) \\ \vdots & \vdots & \vdots\vdots\vdots & \vdots \\ f_p(m_1) & f_p(m_2) & \cdots & f_p(m_p) \end{vmatrix} \end{aligned}$$

because, by the inductive hypothesis, $g(m_1 \wedge \cdots \wedge \hat{m}_i \wedge \cdots \wedge m_p)$ is equal to the determinant obtained by striking out the first row and the i-th column in (7.11.4).

Let us now put

$$G(M) = \sum_{n \in \mathbb{Z}} E_n(M)^*, \tag{7.11.5}$$

where the sum is taken in $E(M)^*$, and

$$G_n(M)=E_n(M)^*. \tag{7.11.6}$$

Then, by Theorem 8, $G(M)$ is an R-subalgebra of $E(M)^*$ and it has $\{G_n(M)\}_{n\in\mathbb{Z}}$ as an algebra-grading. Since

$$G_0(M)=\mathrm{Hom}_R(E_0(M), R)=\mathrm{Hom}_R(R, R), \tag{7.11.7}$$

$G_0(M)$ is a free R-module of rank one. Note that ε is the identity element of $G(M)$ and that it forms a base for $G_0(M)$. We also have

$$G_1(M)=\mathrm{Hom}_R(E_1(M), R)=M^*. \tag{7.11.8}$$

Finally, we note that *if M is a finitely generated R-module, then $G(M)=E(M)^*$*. This is because finite generation ensures that $E_n(M)=0$ for all large values of n.

Definition. *The graded R-algebra $G(M)$ is called the 'Grassmann algebra' of M.*

Theorem 13. *The Grassmann algebra $G(M)$ is an anticommutative algebra.*

Proof. By Lemma 8, $E(M)$ is a skew-commutative coalgebra and therefore, by Theorem 8 Cor., $G(M)$ is a skew-commutative algebra. Now suppose that $f\in G_p(M)$, where p is an *odd* integer. Then, to complete the proof, we need only show that $f\wedge f=0$. Suppose therefore that $m_1, m_2, \ldots, m_{2p}$ belong to M. Since $f\wedge f$ belongs to $G_{2p}(M)$, the theorem will follow if we establish that

$$(f\wedge f)(m_1\wedge m_2\wedge\cdots\wedge m_{2p})=0.$$

But, by Lemma 10,

$$(f\wedge f)(m_1\wedge m_2\wedge\cdots\wedge m_{2p})=\sum \mathrm{sgn}(I, I')f(m_I)f(m_{I'}),$$

where (because $f\in G_p(M)$) we have only to sum over all sequences $I=(i_1, i_2, \ldots, i_p)$ of length p with $1\leq i_1<\cdots<i_p\leq 2p$. (Here I' is obtained by deleting the terms in I from $(1, 2, \ldots, 2p)$.) But, for such an I, $\mathrm{sgn}(I, I')=-\mathrm{sgn}(I', I)$ because p is odd, and therefore

$$(f\wedge f)(m_1\wedge m_2\wedge\cdots\wedge m_{2p})=0$$

as required.

7.12 $S(M)$ as a Hopf algebra

Let M be an R-module and let $S(M)$ be its symmetric algebra. We propose to show how the structure of $S(M)$ as an algebra can be complemented so that the final result is a graded Hopf algebra. In what follows we identify $S_0(M)$ with R.

Let $\varepsilon\colon S(M)\to R$ be the projection of $S(M)$ onto $S_0(M)$. This will be the counit mapping in the coalgebra structure. Note that ε is a homomorphism of R-algebras.

To define the comultiplication or *diagonalization* mapping on $S(M)$, we first observe that there is an R-linear mapping $M \to S(M) \otimes S(M)$ in which m, in M, becomes $m \otimes 1 + 1 \otimes m$. However, $S(M) \otimes S(M)$ is a commutative algebra and therefore the R-linear mapping can be extended to a homomorphism

$$\Delta\colon S(M) \to S(M) \otimes S(M) \tag{7.12.1}$$

of algebras. We record that, by construction,

$$\Delta(m) = m \otimes 1 + 1 \otimes m \tag{7.12.2}$$

for all m in M.

Lemma 12. *With the above notation, $(S(M), \Delta, \varepsilon)$ is a coalgebra. It is graded, as a coalgebra, by the symmetric powers $\{S_n(M)\}_{n\in\mathbb{Z}}$ of the module M. The coalgebra $(S(M), \Delta, \varepsilon)$ is commutative.*

Proof. This lemma is the counterpart of Lemma 8 and it can be proved in very much the same way. The details will therefore be omitted.

In the next theorem, μ and η denote the multiplication and unit mappings of the symmetric algebra.

Theorem 14. *Let the notation be as explained above. Then $(\eta, \mu, S(M), \Delta, \varepsilon)$ is a graded Hopf algebra, the grading being given by the symmetric powers $\{S_n(M)\}_{n\in\mathbb{Z}}$ of the module M. As a Hopf algebra, $S(M)$ is commutative.*

Proof. That $S(M)$ is a Hopf algebra follows from the corollary to Lemma 5. It is obviously commutative.

Theorem 15. *Let $f: M \to N$ be a homomorphism of R-modules. Then the mapping $S(f)\colon S(M) \to S(N)$, which was defined in Section (6.2), is a homomorphism of graded Hopf algebras.*

Proof. Theorem 15 corresponds to Theorem 12. In the case of Theorem 12 we described the essential features of the demonstration. Similar considerations apply here.

In Section (7.11) we used the fact that $E(M)$ is a graded coalgebra in order to derive from it the Grassmann algebra of M. In like manner we can use the graded coalgebra $S(M)$ to derive what we might expect to be a new algebra. Let us see what the construction produces.

To this end we consider

$$\mathfrak{U}_n = S_n(M)^* = \mathrm{Hom}_R(S_n(M), R) \tag{7.12.3}$$

as a submodule of $S(M)^* = \mathrm{Hom}_R(S(M), R)$ and put

$$\mathfrak{U} = \sum_{n\in\mathbb{Z}} \mathfrak{U}_n. \tag{7.12.4}$$

Then, by Theorem 8, $\mathfrak{U}$ is an R-algebra which we can regard as being graded by the family $\{\mathfrak{A}_n\}_{n\in\mathbb{Z}}$. Furthermore, the corollary to Theorem 7 shows that the algebra is commutative. For f, g in $\mathfrak{A}$ we shall use fg to denote their product. Then, by (7.7.4),

$$fg = \mu_R \circ (f \otimes g) \circ \Delta, \tag{7.12.5}$$

where Δ is the diagonalization mapping of $S(M)$.

Next, we recall that in Section (6.7) we defined an algebra $\nabla S(M)$ which we named the algebra of differential operators. By construction it is a commutative subalgebra of $\mathrm{End}_R(S(M))$ and, moreover, it is graded. The module of differential operators of degree h was denoted by $\nabla_h S(M)$.

Suppose that $\phi \in \nabla_h S(M)$, where $h \geq 0$. Then ϕ is an R-endomorphism of $S(M)$ of degree $-h$ and it induces a linear form, $\bar{\phi}$ say, on $S_h(M)$. Also, by Theorem 8 of Chapter 6, the mapping $\phi \mapsto \bar{\phi}$ is an isomorphism of $\nabla_h S(M)$ onto $S_h(M)^*$. Accordingly we have an isomorphism

$$\nabla_h S(M) \xrightarrow{\sim} \mathfrak{A}_h \tag{7.12.6}$$

for each value of h and, by combining these isomorphisms, we arrive at a degree-preserving isomorphism

$$\nabla S(M) \xrightarrow{\sim} \mathfrak{A} \tag{7.12.7}$$

of R-modules.

Theorem 16. *The module-isomorphism* (7.12.7) *is actually a degree-preserving algebra-isomorphism of the algebra* $\nabla S(M)$ *of differential operators onto the algebra* $\mathfrak{A}$.

Proof. Suppose that $\phi \in \nabla_h S(M)$ and $\psi \in \nabla_k S(M)$, where $h \geq 0$ and $k \geq 0$. Then $\phi \circ \psi$ is a differential operator of degree $h+k$ and its image in $\mathfrak{A}_{h+k}$ is its restriction $\overline{\phi \circ \psi}$ to $S_{h+k}(M)$. The theorem will follow if we show that $\overline{\phi \circ \psi}$ is the product of $\bar{\phi}$ and $\bar{\psi}$ in $\mathfrak{A}$.

To see this let $m_1, m_2, \ldots, m_{h+k}$ belong to M. It will now suffice to prove that $(\phi \circ \psi)(m_1 m_2 \ldots m_{h+k})$ and $(\bar{\phi}\bar{\psi})(m_1 m_2 \ldots m_{h+k})$ are the same. In this connection we note that

$$\begin{aligned}(\phi \circ \psi)&(m_1 m_2 \ldots m_{h+k})\\ &= \sum \phi(m_{i_1} m_{i_2} \ldots m_{i_h})\psi(m_{j_1} m_{j_2} \ldots m_{j_k}),\end{aligned}$$

where $(i_1, i_2, \ldots, i_h)$ and $(j_1, j_2, \cdots, j_k)$ constitute a variable pair of strictly increasing sequences of integers which, between them, include all the positive whole numbers from 1 to $h+k$ (see (6.7.6)). Next, from (7.7.4), we see that $\bar{\phi}\bar{\psi} = \mu_R \circ (\bar{\phi} \otimes \bar{\psi}) \circ \Delta$, where in this context $\bar{\phi}$ and $\bar{\psi}$ are regarded as linear forms on $S(M)$ and Δ is the diagonalization mapping of the symmetric algebra. Furthermore $\Delta(m_1 m_2 \ldots m_{h+k})$ is the product, in

$S(M) \otimes S(M)$, of the elements

$$m_i \otimes 1 + 1 \otimes m_i \quad (i=1,2,\ldots,h+k).$$

Of course, these elements are of degree one and $S(M) \otimes S(M)$ is a commutative algebra; also $\bar{\phi}$ is null on $S_n(M)$ if $n \neq h$, whereas $\bar{\psi}$ is null on $S_n(M)$ if $n \neq k$. From these observations we conclude that

$$((\bar{\phi} \otimes \bar{\psi}) \circ \Delta)(m_1 m_2 \ldots m_{h+k})$$
$$= \sum \bar{\phi}(m_{i_1} m_{i_2} \ldots m_{i_h}) \otimes \bar{\psi}(m_{j_1} m_{j_2} \ldots m_{j_k}),$$

where the sum is taken over the same pairs of sequences as before. But ϕ extends $\bar{\phi}$ and ψ extends $\bar{\psi}$. Consequently

$$(\bar{\phi}\bar{\psi})(m_1 m_2 \ldots m_{h+k})$$
$$= \sum \phi(m_{i_1} m_{i_2} \ldots m_{i_h}) \psi(m_{j_1} m_{j_2} \ldots m_{j_k})$$

and with this the proof is complete.

Thus, to sum up, when we consider linear forms on the coalgebra $S(M)$ we do not obtain an essentially new graded algebra; instead we recover up to isomorphism the algebra of differential operators, on $S(M)$, that we encountered in Chapter 6.

7.13 Comments and exercises

It will be clear to the reader, and indeed it has been mentioned already, that the analogy between the theory of algebras and the theory of coalgebras can be developed further than was done in the main text. Thus our first exercise corresponds to the fact that the tensor product of a number of commutative algebras is itself commutative, and the second exercise is suggested by Theorem 7 of Chapter 3. The two exercises can be solved in very much the same way, so a solution will be provided only for the second one. As on earlier occasions, those exercises for which a solution is given are marked with an asterisk.

Exercise 1. *Show that if $A_1, A_2, \ldots, A_p$ are commutative R-coalgebras, then $A_1 \otimes A_2 \otimes \cdots \otimes A_p$ is also a commutative R-coalgebra.*

Exercise 2*. *Show that if $A^{(1)}, A^{(2)}, \ldots, A^{(p)}$ are skew-commutative R-coalgebras, then $A^{(1)} \underline{\otimes} A^{(2)} \underline{\otimes} \cdots \underline{\otimes} A^{(p)}$ is also a skew-commutative R-coalgebra.*

The next exercise corresponds to Exercise 6 of Chapter 3 and it may be solved in much the same way provided that Theorem 6 (of this chapter) is used to take over the role previously played by Exercise 5 of Chapter 3.

Exercise 3. *Let $A^{(1)}, A^{(2)}, \ldots, A^{(p)}$ be graded R-coalgebras. Show that if the factors in $A^{(1)} \underline{\otimes} A^{(2)} \underline{\otimes} \cdots \underline{\otimes} A^{(p)}$ are permuted in any way, then the* *modified tensor product remains unchanged to within an isomorphism of graded coalgebras.*

The formal difference between the definition of an *anticommutative* algebra and that of a *skew-commutative* algebra is clear enough. The next exercise invites the reader to construct an example to show that the distinction is a real one.

Exercise 4*. *Give an example of a skew-commutative algebra which is not anticommutative.*

Next we return to the *original* definition of a graded algebra. Suppose then that A is an R-algebra and $\{A_n\}_{n\in\mathbb{Z}}$ is a family of R-submodules of A that grades A as a *module.* By the definition given in Section (3.3), $\{A_n\}_{n\in\mathbb{Z}}$ grades A as an *algebra* if from $a_r \in A_r$ and $a_s \in A_s$ it follows that $a_r a_s$ belongs to A_{r+s}. Thus we have an algebra-grading provided only that the multiplication mapping $\mu: A \otimes A \to A$ preserves degrees. However, by Theorem 5 of Chapter 3, in these circumstances the identity element of A is homogeneous of degree zero and so the unit mapping $\eta: R \to A$ automatically preserves degrees as well. The following exercise shows that a similar situation exists in the case of coalgebras.

Exercise 5*. *Let A be an R-coalgebra and let $\{A_n\}_{n\in\mathbb{Z}}$ be a family of R-submodules of A such that*

$$A = \sum_{n\in\mathbb{Z}} A_n \quad \text{(d.s.).}$$

Further, let the comultiplication mapping $\Delta: A \to A \otimes A$ preserve the degrees of homogeneous elements. Show that if R is given the trivial grading, then the counit mapping $\varepsilon: A \to R$ also preserves degrees.

We now add some comments concerning the results of Section (7.7). There we saw that if A is an R-coalgebra and $A^* = \mathrm{Hom}_R(A, R)$, then A^* has a natural structure as an R-algebra. Let us examine the functorial aspects of the way the algebra A^* depends on the coalgebra A.

First we note that each R-linear mapping $\phi: A \to B$, of A into an R-module B, gives rise to a module-homomorphism

$$\phi^*: B^* \to A^* \tag{7.13.1}$$

in which $\phi^*(f) = f \circ \phi$.

Exercise 6*. *If $\phi: A \to B$ is a homomorphism of R-coalgebras, show that $\phi^*: B^* \to A^*$ is a homomorphism of R-algebras.*

In fact the association of A^* with A provides an example of what is known as a *contravariant* functor. This particular functor is from R-coalgebras to R-algebras.

Let M be an R-module. In Theorem 16 we saw that when $S(M)$ is regarded as a graded coalgebra and we use Theorem 8 to derive from it a graded algebra, what we obtain is isomorphic to the algebra of differential operators on $S(M)$; in particular the application of Theorem 8 produces an algebra that is isomorphic to a certain graded subalgebra of $\mathrm{End}_R(S(M))$. This suggests that the Grassmann algebra $G(M)$ may be related to some graded subalgebra of $\mathrm{End}_R(E(M))$. As will now be shown this is indeed the case.

Suppose that $h \geq 0$ and consider the R-endomorphisms of $E(M)$ of degree $-h$. Such an endomorphism induces a linear form on $E_h(M)$. In what follows p denotes an integer satisfying $p \geq h$, $I = (i_1, i_2, \ldots, i_h)$ is a sequence of integers with $1 \leq i_1 < i_2 < \cdots < i_h \leq p$, and I' denotes the increasing sequence obtained from $(1, 2, \ldots, p)$ by deleting the terms that belong to I. Finally, if $m_1, m_2, \ldots, m_p$ belong to M, then m_I will be used for the element $m_{i_1} \wedge m_{i_2} \wedge \cdots \wedge m_{i_p}$ and $m_{I'}$ is to be defined similarly.

By a *differential form* on $E(M)$ of degree h, we shall understand an R-endomorphism ϕ of degree $-h$ that has the following property: for all $p \geq h$ and all $m_1, m_2, \ldots, m_p$ in M

$$\phi(m_1 \wedge m_2 \wedge \cdots \wedge m_p) = \sum_I \mathrm{sgn}(I, I')\phi(m_I)m_{I'}. \tag{7.13.2}$$

As usual, if $I = (i_1, i_2, \ldots)$ and $I' = (i'_1, i'_2, \ldots)$, then $\mathrm{sgn}(I, I')$ stands for the sign of $(i_1, i_2, \ldots, i'_1, i'_2, \ldots)$ considered as a permutation of $(1, 2, \ldots, p)$.

Evidently the differential forms of degree h constitute an R-submodule of $\mathrm{End}_R(E(M))$; and if two differential forms of degree h induce the same linear form on $E_h(M)$, then they coincide.

Exercise 7*. *Let ϕ and ψ be differential forms on $E(M)$ of degrees $h \geq 0$ and $k \geq 0$ respectively. Show that $\psi \circ \phi$ and $\phi \circ \psi$ are differential forms of degree $h+k$ and that $\psi \circ \phi = (-1)^{hk}\phi \circ \psi$. Show also that if h is odd, then $\phi \circ \phi = 0$.*

The next exercise should be compared with Theorem 8 of Chapter 6.

Exercise 8*. *Let M be an R-module and f a linear form on $E_h(M)$, where $h \geq 0$. Show that, on $E(M)$, there is exactly one differential form of degree h that extends f.*

From Exercises 7 and 8, we see that there is a subalgebra of $\mathrm{End}_R(E(M))$ that is non-negatively graded by the modules of differential forms. Let us call this the *algebra of differential forms* on $E(M)$. By Exercise 7, the algebra is anticommutative.

Let ϕ be a differential form on $E(M)$ and let the degree of ϕ be h. Then ϕ restricts to a linear form, $\bar{\phi}$ say, on $E_h(M)$. Note that $\bar{\phi}$ belongs to the Grassmann algebra $G(M)$; in fact $\bar{\phi} \in G_h(M)$.

Exercise 9*. *Let ϕ and ψ be differential forms on $E(M)$. Show that the linear form $\overline{\psi \circ \phi}$ associated with $\psi \circ \phi$ satisfies $\overline{\psi \circ \phi} = \bar{\phi} \wedge \bar{\psi}$, where $\bar{\phi} \wedge \bar{\psi}$ is the product of $\bar{\phi}$ and $\bar{\psi}$ in $G(M)$.*

Let A be a ring with identity and let $\bar{A}$ be the ring that is *opposite* to A. (Thus A and $\bar{A}$ coincide as sets and addition is the same for both of them; however, the product of two elements of $\bar{A}$ is their product in A but with the order of the factors reversed.) If A happens to be an R-algebra, then $\bar{A}$ will also be an R-algebra with the same R-module structure as A. Naturally we refer to $\bar{A}$ as the *opposite algebra* to A. Finally, if $\{A_n\}_{n \in \mathbb{Z}}$ grades A as an R-algebra, then the same family provides an algebra-grading on the opposite algebra.

This terminology enables us to sum up our main conclusions as follows: *the Grassmann algebra $G(M)$ and the opposite of the algebra of differential forms on $E(M)$ are isomorphic graded algebras.*

7.14 Solutions to selected exercises

Exercise 2. *Show that if $A^{(1)}, A^{(2)}, \ldots, A^{(p)}$ are skew-commutative R-coalgebras, then $A^{(1)} \underline{\otimes} A^{(2)} \underline{\otimes} \cdots \underline{\otimes} A^{(p)}$ is also a skew-commutative R-coalgebra.*

Solution. The property of being skew-commutative is preserved under an isomorphism of graded coalgebras, and because of this it is enough to prove the result in question in the case of *two* coalgebras. We shall therefore assume that A and B are skew-commutative coalgebras and prove that $A \underline{\otimes} B$ is skew-commutative as well.

We use our usual notation. Let a_m belong to A_m and b_n to B_n. Further let $\Delta_A(a_m) = \sum a' \otimes a''$ and $\Delta_B(b_n) = \sum b' \otimes b''$, where a', a'' are homogeneous elements of A and b', b'' homogeneous elements of B. Then

$$\Delta_{A \underline{\otimes} B}(a_m \otimes b_n) = \sum (-1)^{\{a''\}\{b'\}} (a' \otimes b') \otimes (a'' \otimes b''),$$

where it is understood that $\{a'\}, \{a''\}$ denote the degrees of a' and a'', and similarly in the case of $\{b'\}$ and $\{b''\}$.

Let $T_A: A \otimes A \xrightarrow{\sim} A \otimes A$, $T_B: B \otimes B \xrightarrow{\sim} B \otimes B$ and

$$T: (A \underline{\otimes} B) \otimes (A \underline{\otimes} B) \xrightarrow{\sim} (A \underline{\otimes} B) \otimes (A \underline{\otimes} B)$$

be twisting isomorphisms. Then

$$(T \circ \Delta_{A \underline{\otimes} B})(a_m \otimes b_n)$$
$$= \sum (-1)^{\{a'\}\{a''\} + \{b'\}\{b''\}} (-1)^{\{a'\}\{b''\}} (a'' \otimes b'') \otimes (a' \otimes b').$$

But $T_A \circ \Delta_A = \Delta_A$ and $T_B \circ \Delta_B = \Delta_B$. Consequently

$$\Delta_A(a_m) = \sum (-1)^{\{a'\}\{a''\}} (a'' \otimes a'),$$
$$\Delta_B(b_n) = \sum (-1)^{\{b'\}\{b''\}} (b'' \otimes b')$$

and now it follows that

$$\begin{aligned}&\Delta_{A\otimes B}(a_m\otimes b_n)\\&=\sum(-1)^{\{a'\}\{a''\}+\{b'\}\{b''\}}(-1)^{\{a'\}\{b''\}}(a''\otimes b'')\otimes(a'\otimes b')\\&=(T\circ\Delta_{A\otimes B})(a_m\otimes b_n).\end{aligned}$$

Accordingly $T\circ\Delta_{A\otimes B}=\Delta_{A\otimes B}$ and with this the solution is complete.

Exercise 4. *Give an example of a skew-commutative algebra which is not anticommutative.*

Solution. Let F be a field of characteristic two and let A be the free F-algebra generated by a set $\{X, Y\}$ of two elements. Then $XY+YX$ generates a homogeneous two-sided ideal, I say, in A. Let B denote the graded F-algebra A/I and denote by x and y the natural images of X and Y in B.

The homogeneous elements of B that have degree one form the vector space $B_1=Fx+Fy$, and this generates B as an algebra. By construction $xy+yx=0$ and now, because F has characteristic two, we have $bb'+b'b=0$ for all b, b' in B_1. That B is skew-commutative is a simple consequence of this.

Finally $X^2\notin I$ and therefore $x^2\neq 0$. This shows that B is not anticommutative.

Exercise 5. *Let A be an R-coalgebra and let $\{A_n\}_{n\in\mathbb{Z}}$ be a family of R-submodules of A such that*

$$A=\sum_{n\in\mathbb{Z}} A_n \quad \text{(d.s.)}.$$

Further, let the comultiplication mapping $\Delta\colon A\to A\otimes A$ preserve the degrees of homogeneous elements. Show that if R is given the trivial grading, then the counit mapping $\varepsilon\colon A\to R$ also preserves degrees.

Solution. Suppose that $a_m\in A_m$, where $m\neq 0$. It is sufficient to show that $\varepsilon(a_m)=0$.

We can write $\Delta(a_m)=\sum\alpha_i\otimes\beta_i$, where α_i, β_i are homogeneous elements of A the sum of whose degrees is m. Now, because ε is the counit mapping, we have

$$a_m=\sum\varepsilon(\alpha_i)\beta_i=\sum\varepsilon(\beta_i)\alpha_i \tag{7.14.1}$$

and therefore $\varepsilon(a_m)=\sum\varepsilon(\alpha_i)\varepsilon(\beta_i)$.

Let us use $\{\alpha_i\}$, respectively $\{\beta_i\}$, to denote the degree of α_i, respectively β_i. Then $\{\alpha_i\}\neq 0$ if and only if $\{\beta_i\}\neq m$. It follows, from (7.14.1), that

$$\sum_{i:\{\alpha_i\}\neq 0}\varepsilon(\alpha_i)\beta_i=0$$

and therefore

$$\sum_{i:\{\alpha_i\}\neq 0} \varepsilon(\alpha_i)\varepsilon(\beta_i)=0. \tag{7.14.2}$$

But, again from (7.14.1),

$$\sum_{i:\{\alpha_i\}=0} \varepsilon(\beta_i)\alpha_i=0$$

and so

$$\sum_{i:\{\alpha_i\}=0} \varepsilon(\alpha_i)\varepsilon(\beta_i)=0.$$

If now we combine this with (7.14.2) we find that $\sum \varepsilon(\alpha_i)\varepsilon(\beta_i)=0$, that is to say $\varepsilon(a_m)=0$. This completes the solution.

Exercise 6. *If $\phi: A \to B$ is a homomorphism of R-coalgebras, show that $\phi^*: B^* \to A^*$ is a homomorphism of R-algebras.*

Solution. With the usual notation we have $\phi^*(\varepsilon_B)=\varepsilon_B \circ \phi = \varepsilon_A$ because ϕ preserves counits. Consequently ϕ^* preserves identity elements.

Now suppose that h_1 and h_2 belong to B^*. By (7.7.4), their product in this algebra is $\mu_R \circ (h_1 \otimes h_2) \circ \Delta_B$ and, moreover,

$$\begin{aligned}\phi^*(\mu_R \circ (h_1 \otimes h_2) \circ \Delta_B) &= \mu_R \circ (h_1 \otimes h_2) \circ \Delta_B \circ \phi \\ &= \mu_R \circ (h_1 \otimes h_2) \circ (\phi \otimes \phi) \circ \Delta_A\end{aligned}$$

because ϕ is compatible with comultiplication. But

$$\begin{aligned}\mu_R \circ (h_1 \otimes h_2) \circ (\phi \otimes \phi) \circ \Delta_A &= \mu_R \circ (h_1 \circ \phi) \otimes (h_2 \circ \phi) \circ \Delta_A \\ &= \mu_R \circ \phi^*(h_1) \otimes \phi^*(h_2) \circ \Delta_A\end{aligned}$$

and this is just the product of $\phi^*(h_1)$ and $\phi^*(h_2)$ in A^*. Thus ϕ^* is compatible with multiplication and the solution is complete.

Exercise 7. *Let ϕ and ψ be differential forms on $E(M)$ of degrees $h \geq 0$ and $k \geq 0$ respectively. Show that $\psi \circ \phi$ and $\phi \circ \psi$ are differential forms of degree $h+k$ and that $\psi \circ \phi = (-1)^{hk} \phi \circ \psi$. Show also that if h is odd, then $\phi \circ \phi = 0$.*

Solution. Evidently $\psi \circ \phi$ and $\phi \circ \psi$ are endomorphisms of $E(M)$ and each has degree $-(h+k)$. Now suppose that $p \geq h+k$ and that $m_1, m_2, \ldots, m_p$ belong to M. In what follows $I=(i_1, i_2, \ldots, i_h)$ and $J=(j_1, j_2, \ldots, j_k)$ will denote a pair of sequences of integers, where $1 \leq i_1 < \cdots < i_h \leq p$, $1 \leq j_1 < \cdots < j_k \leq p$ and I and J *have no term in common.* We put $m_I = m_{i_1} \wedge m_{i_2} \wedge \cdots \wedge m_{i_h}$ and define m_J similarly. Further we use I', respectively J', to denote the sequence obtained from $(1, 2, \ldots, p)$ by striking out the terms that belong to I, respectively J; and finally $(IJ)'$ will designate the sequence obtained, from $(1, 2, \ldots, p)$, when the terms of both I and J are removed.

By (7.13.2),

$$\phi(m_1 \wedge m_2 \wedge \cdots \wedge m_p) = \sum_I \operatorname{sgn}(I, I')\phi(m_I)m_{I'}$$

and now if we operate on this with ψ we obtain

$$(\psi \circ \phi)(m_1 \wedge m_2 \wedge \cdots \wedge m_p)$$
$$= \sum_{I,J} \operatorname{sgn}(I, I') \operatorname{sgn}(J, (IJ)')\phi(m_I)\psi(m_J)m_{(IJ)'}.$$

Note that the sequence J followed by the sequence $(IJ)'$ is a permutation of I'. It is the sign of this permutation that is denoted by $\operatorname{sgn}(J, (IJ)')$.

Now I, J and $(IJ)'$, taken in order, provide a permutation of $(1, 2, \ldots, p)$. If $\operatorname{sgn}(I, J, (IJ)')$ denotes the sign of this permutation, then

$$\operatorname{sgn}(I, I') \operatorname{sgn}(J, (IJ)') = \operatorname{sgn}(I, J, (IJ)')$$

and therefore

$$(\psi \circ \phi)(m_1 \wedge m_2 \wedge \cdots \wedge m_p)$$
$$= \sum_{I,J} \operatorname{sgn}(I, J, (IJ)')\phi(m_I)\psi(m_J)m_{(IJ)'}. \tag{7.14.3}$$

Similarly we find that

$$(\phi \circ \psi)(m_1 \wedge m_2 \wedge \cdots \wedge m_p)$$
$$= \sum_{J,I} \operatorname{sgn}(J, I, (JI)')\psi(m_J)\phi(m_I)m_{(JI)'} \tag{7.14.4}$$

and from (7.14.3) and (7.14.4) it is clear that $\psi \circ \phi = (-1)^{hk}\phi \circ \psi$.

For the last step we assume that h is odd and we take $\psi = \phi$ so that $h = k$. By (7.14.3),

$$(\phi \circ \phi)(m_1 \wedge m_2 \wedge \cdots \wedge m_p) = \sum_{I,J} \operatorname{sgn}(I, J, (IJ)')\phi(m_I)\psi(m_J)m_{(IJ)'}$$
$$= 0$$

because, under the present assumptions,

$$\operatorname{sgn}(I, J, (IJ)') + \operatorname{sgn}(J, I, (JI)') = 0.$$

Accordingly $\phi \circ \phi = 0$ as required.

Exercise 8. *Let M be an R-module and f a linear form on $E_h(M)$, where $h \geq 0$. Show that, on $E(M)$, there is exactly one differential form of degree h that extends f.*

Solution. Suppose that $p \geq h$ and define a mapping of the p-fold product $M \times M \times \cdots \times M$ into $E_{p-h}(M)$ by

$$(m_1,m_2,\ldots,m_p)\mapsto\sum_I \operatorname{sgn}(I,I')f(m_I)m_{I'},$$

where the notation is the same as in (7.13.2). This mapping is clearly multilinear.

We claim that it is also an alternating mapping. For suppose that $1\leq j<p$ and that $m_j=m_{j+1}$. If both j and $j+1$ are in I, then $m_I=0$; and if neither is in I then $m_{I'}=0$. Thus in both these cases $f(m_I)m_{I'}=0$.

The remaining I's are those which contain just one of j and $j+1$. Such sequences can be arranged in pairs, I_1 and I_2 say, where each member of a pair becomes the other member when j and $j+1$ are interchanged. Now $f(m_{I_1})m_{I_1'}$ is equal to $f(m_{I_2})m_{I_2'}$, because $m_j=m_{j+1}$; and

$$\operatorname{sgn}(I_1,I_1')=-\operatorname{sgn}(I_2,I_2').$$

Consequently

$$\operatorname{sgn}(I_1,I_1')f(m_{I_1})m_{I_1'}+\operatorname{sgn}(I_2,I_2')f(m_{I_2})m_{I_2'}=0$$

and therefore we have

$$\sum_I \operatorname{sgn}(I,I')f(m_I)m_{I'}=0.$$

This establishes our claim (see Chapter 1, Lemma 2).

It now follows that there is an R-linear mapping of $E_p(M)$ into $E_{p-h}(M)$ in which

$$m_1\wedge m_2\wedge\cdots\wedge m_p\mapsto\sum_I \operatorname{sgn}(I,I')f(m_I)m_{I'}.$$

When $p=h$ this is just f. Accordingly if we combine these various homomorphisms into an endomorphism of $E(M)$ it will extend f; moreover, by construction, the endomorphism so obtained will be a differential form of degree h. This provides all that is needed because the part of the exercise that is concerned with uniqueness has been dealt with already.

Exercise 9. *Let ϕ and ψ be differential forms on $E(M)$. Show that the linear form $\overline{\psi\circ\phi}$ associated with $\psi\circ\phi$ satisfies $\overline{\psi\circ\phi}=\bar{\phi}\wedge\bar{\psi}$, where $\bar{\phi}\wedge\bar{\psi}$ is the product of $\bar{\phi}$ and $\bar{\psi}$ in $G(M)$.*

Solution. Let the degrees of ϕ and ψ be $h\geq 0$ and $k\geq 0$ respectively, and let $m_1, m_2, \ldots, m_{h+k}$ belong to M. By (7.13.2),

$$\phi(m_1\wedge m_2\wedge\cdots\wedge m_{h+k})=\sum_I \operatorname{sgn}(I,I')\phi(m_I)m_{I'}$$

and therefore

$$(\psi\circ\phi)(m_1\wedge m_2\wedge\cdots\wedge m_{h+k})=\sum_I \operatorname{sgn}(I,I')\phi(m_I)\psi(m_{I'}).$$

Here $I=(i_1, i_2, \ldots, i_h)$ is an increasing sequence of integers between 1 and $h+k$ and I' denotes the residual sequence.

We now consider $(\bar{\phi} \wedge \bar{\psi})(m_1 \wedge m_2 \wedge \cdots \wedge m_{h+k})$. This can be obtained from Lemma 10; and, if we remember that $\bar{\phi}$ and $\bar{\psi}$ (considered as elements of $E(M)^*$) belong to $E_h(M)^*$ and $E_k(M)^*$ respectively, we find that the relation provided by the lemma is

$$(\bar{\phi} \wedge \bar{\psi})(m_1 \wedge m_2 \wedge \cdots \wedge m_{h+k}) = \sum_I \operatorname{sgn}(I, I')\bar{\phi}(m_I)\bar{\psi}(m_{I'}),$$

where I ranges over the same set of sequences as before. However, $\bar{\phi}(m_I) = \phi(m_I)$ and $\bar{\psi}(m_{I'}) = \psi(m_{I'})$, and therefore we have $\overline{\psi \circ \phi} = \bar{\phi} \wedge \bar{\psi}$ as required.

8

Graded duality

General remarks

From a graded coalgebra it is possible, by considering linear forms on the submodules which make up the grading, to derive a graded algebra. In fact, this is the conclusion of Theorem 8 of Chapter 7. An attempt to perform a similar construction starting instead with a general graded algebra soon runs into difficulties; but when the algebra is suitably restricted the process can be carried through successfully to produce a graded coalgebra as the end-product.

In the present chapter these ideas are developed in detail and it is shown that, for non-negatively graded algebras and coalgebras whose grading modules are free modules of finite rank, one has a full duality theory. This not only interchanges algebras and coalgebras, but also preserves ordinary and modified Hopf algebras.

The first two sections of the chapter are used to establish the results on modules that provide the basis of this theory. As always, R denotes a commutative ring with an identity element, and the symbol $\otimes$, when used without a subscript, indicates a tensor product formed over R.

8.1 Modules of linear forms

It was remarked in the introduction to this chapter that the duality we shall be discussing arises from the study of linear forms on free modules with finite bases. However, before we restrict our attention to such modules it is convenient to make some observations concerning linear forms on arbitrary modules.

Suppose then that M and N are R-modules and put

$$M^* = \mathrm{Hom}_R(M, R), \tag{8.1.1}$$

$N^* = \mathrm{Hom}_R(N, R)$ so that M^* and N^* are the so-called *duals* of M and N

respectively. If now $\lambda: M \to N$ is R-linear, then the *dual homomorphism*

$$\lambda^*: N^* \to M^* \tag{8.1.2}$$

is defined by

$$\lambda^*(\phi) = \phi \circ \lambda \tag{8.1.3}$$

for all ϕ in N^*. Note that if id_M denotes the identity mapping of M, then

$$(\mathrm{id}_M)^* = \mathrm{id}_{M^*}; \tag{8.1.4}$$

and that if we have a second homomorphism, $\mu: N \to K$ say, then

$$(\mu \circ \lambda)^* = \lambda^* \circ \mu^*. \tag{8.1.5}$$

(In the language of Category Theory, the taking of duals is a *contravariant functor* from R-modules to R-modules.)

If we consider R as a module and form its dual R^*, then this is isomorphic to R itself; in fact we have an isomorphism

$$R^* \approx R \tag{8.1.6}$$

in which f in R^* is matched with $f(1)$ in R. This will be called the *canonical isomorphism* of R^* onto R and, in the sequel, it will often be used to identify R^* with R.

An R-module M is connected with its double dual M^{**} by means of an R-linear mapping

$$M \to M^{**} \tag{8.1.7}$$

in which m in M becomes the homomorphism $M^* \to R$ given by $f \mapsto f(m)$. The homomorphism $M \to M^{**}$ is *natural* in the sense that if $\lambda: M \to N$ is R-linear, then

$$\begin{array}{ccc} M & \longrightarrow & M^{**} \\ {\scriptstyle\lambda}\downarrow & & \downarrow{\scriptstyle\lambda^{**}} \\ N & \longrightarrow & N^{**} \end{array} \tag{8.1.8}$$

is a commutative diagram.

We must now consider the duals of finite direct sums and of tensor products. The former are easily described for if $M_1, M_2, \ldots, M_p$ are R-modules, then we have an isomorphism

$$(M_1 \oplus M_2 \oplus \cdots \oplus M_p)^* \approx M_1^* \oplus M_2^* \oplus \cdots \oplus M_p^*; \tag{8.1.9}$$

here f in $(M_1 \oplus M_2 \oplus \cdots \oplus M_p)^*$ corresponds to $(f_1, f_2, \ldots, f_p)$, where f_i is the restriction of f to M_i.

For tensor products the situation is more complicated. Let us again assume that $M_1, M_2, \ldots, M_p$ are R-modules and let f_i belong to M_i^* for $i = 1, 2, \ldots, p$. If now

$$\mu_R^{(p)}: R \otimes R \otimes \cdots \otimes R \to R \tag{8.1.10}$$

denotes p-fold multiplication (so that $\mu_R^{(p)}(r_1 \otimes r_2 \otimes \cdots \otimes r_p)$ is equal to $r_1 r_2 \ldots r_p$), then $\mu_R^{(p)} \circ (f_1 \otimes f_2 \otimes \cdots \otimes f_p)$ is a linear form on $M_1 \otimes M_2 \otimes \cdots \otimes M_p$ and the mapping

$$M_1 \times M_2 \times \cdots \times M_p \to (M_1 \otimes M_2 \otimes \cdots \otimes M_p)^*$$

which takes $(f_1, f_2, \ldots, f_p)$ into $\mu_R^{(p)} \circ (f_1 \otimes f_2 \otimes \cdots \otimes f_p)$ is multilinear. Accordingly there is an R-linear mapping

$$\omega: M_1^* \otimes M_2^* \otimes \cdots \otimes M_p^* \to (M_1 \otimes M_2 \otimes \cdots \otimes M_p)^* \tag{8.1.11}$$

which is such that

$$\omega(f_1 \otimes f_2 \otimes \cdots \otimes f_p) = \mu_R^{(p)} \circ (f_1 \otimes f_2 \otimes \cdots \otimes f_p) \tag{8.1.12}$$

and therefore, when $m_i \in M_i$ for $i = 1, 2, \ldots, p$,

$$\omega(f_1 \otimes f_2 \otimes \cdots \otimes f_p)(m_1 \otimes m_2 \otimes \cdots \otimes m_p)$$
$$= f_1(m_1) f_2(m_2) \ldots f_p(m_p). \tag{8.1.13}$$

It will be noticed that there is an ambiguity surrounding the use of $f_1 \otimes f_2 \otimes \cdots \otimes f_p$. On the one hand it can denote an element of $M_1^* \otimes M_2^* \otimes \cdots \otimes M_p^*$ and on the other it can stand for a mapping of $M_1 \otimes M_2 \otimes \cdots \otimes M_p$ into $R \otimes R \otimes \cdots \otimes R$; indeed in (8.1.12) both interpretations occur in the same formula. The reader will therefore need to examine the context of each occurrence to discover the appropriate meaning.

It will be seen shortly that when we restrict our attention to free modules with finite bases the homomorphisms (8.1.7) and (8.1.11) become isomorphisms. To simplify the language used to describe these matters, we shall employ the term *finite free* module to describe a free module of finite rank.

Let M be such an R-module and let $b_1, b_2, \ldots, b_s$ be one of its bases. Then for each integer i satisfying $1 \le i \le s$ there is a unique f_i in M^* such that

$$f_i(b_j) = \begin{cases} 0 & \text{if } i \neq j, \\ 1 & \text{if } i = j. \end{cases} \tag{8.1.14}$$

It is easily seen that $f_1, f_2, \ldots, f_s$ is a base of M^* and therefore M^* is a finite free module of the same rank as M itself. The particular base $(f_1, f_2, \ldots, f_s)$ is said to be *dual* to the base $(b_1, b_2, \ldots, b_s)$ of M.

Still supposing that M is a finite free module, let $\theta: M \to M^{**}$ be the homomorphism (8.1.7). Then, using the same notation as in the last paragraph, we find that $\theta(b_i)(f_j) = f_j(b_i)$ has the value 1 if $i = j$ and is zero otherwise. Thus $\theta(b_1), \theta(b_2), \ldots, \theta(b_s)$ is the base of M^{**} that is dual to the base $f_1, f_2, \ldots, f_s$ of M^*.

Lemma 1. *Let M be a finite free R-module. Then the natural homomorphism $M \to M^{**}$ is an isomorphism.*

This is clear because we have just seen that $M \to M^{**}$ takes a base of M into a base of M^{**}. Of course, when M is a finite free module, Lemma 1 may be used to identify M^{**} with M.

Now suppose that $M_1, M_2, \ldots, M_p$ are all of them finite free modules, let B_i be a base of M_i, and let B_i^* be the base of M_i^* that is dual to B_i. In what follows b_i denotes an element of B_i and f_j' an element of B_j^*.

By Theorem 3 of Chapter 1, the products $b_1 \otimes b_2 \otimes \cdots \otimes b_p$ form a base of $M_1 \otimes M_2 \otimes \cdots \otimes M_p$ whereas the products $f_1' \otimes f_2' \otimes \cdots \otimes f_p'$ constitute a base of $M_1^* \otimes M_2^* \otimes \cdots \otimes M_p^*$. If now

$$\omega\colon M_1^* \otimes M_2^* \otimes \cdots \otimes M_p^* \to (M_1 \otimes M_2 \otimes \cdots \otimes M_p)^*$$

is the homomorphism (8.1.11), then, by (8.1.13),

$$\omega(f_1' \otimes f_2' \otimes \cdots \otimes f_p')(b_1 \otimes b_2 \otimes \cdots \otimes b_p)$$

$$= f_1'(b_1) f_2'(b_2) \ldots f_p'(b_p)$$

But this is zero unless

$$f_1'(b_1) = f_2'(b_2) = \cdots = f_p'(b_p) = 1$$

in which case it has the value 1. It follows that $\{\omega(f_1' \otimes f_2' \otimes \cdots \otimes f_p')\}$ is the base of $(M_1 \otimes M_2 \otimes \cdots \otimes M_p)^*$ that is dual to the base $\{b_1 \otimes b_2 \otimes \cdots \otimes b_p\}$ of $M_1 \otimes M_2 \otimes \cdots \otimes M_p$. Accordingly ω maps a base of $M_1^* \otimes M_2^* \otimes \cdots \otimes M_p^*$ into a base of $(M_1 \otimes M_2 \otimes \cdots \otimes M_p)^*$ and we have proved

Lemma 2. *Let $M_1, M_2, \ldots, M_p$ be finite free R-modules. Then the natural homomorphism*

$$M_1^* \otimes M_2^* \otimes \cdots \otimes M_p^* \to (M_1 \otimes M_2 \otimes \cdots \otimes M_p)^*$$

(*see* (8.1.11)) *is an isomorphism.*

This lemma shows how, when $M_1, M_2, \ldots, M_p$ are finite free modules, it is possible to identify $M_1^* \otimes M_2^* \otimes \cdots \otimes M_p^*$ with $(M_1 \otimes M_2 \otimes \cdots \otimes M_p)^*$.

8.2 The graded dual of a graded module

Finite free modules are, of course, very special. More general than these are graded modules with finite free components (the *components* of a graded module are the submodules which make up its grading), and the aim of this section is to extend our results to this wider class. However, we shall begin by considering arbitrary graded modules.

Let M be an R-module and $\{M_n\}_{n \in \mathbb{Z}}$ a grading on M. Then $M_n^* = \mathrm{Hom}_R(M_n, R)$ can be regarded as a submodule of $M^* = \mathrm{Hom}_R(M, R)$; more precisely, if we fix n, then the linear forms on M that vanish on $\sum_{k \neq n} M_k$ constitute a submodule of M^* that is isomorphic to M_n^*. In what follows M_n^*

will be embedded in M^* by means of this isomorphism. The reader will recall that we have already met this kind of situation in Section (7.7).

This said, we can form

$$\sum_n M_n^* = M^\dagger \tag{8.2.1}$$

say, where the summation is carried out in M^*. It is easy to check that the sum in (8.2.1) is direct. Accordingly $M^\dagger$ is graded by $\{M_n^*\}_{n\in\mathbb{Z}}$.

Definition. *The graded module $M^\dagger$ is called the 'graded dual' of the graded module M.*

Now suppose that M and N are graded R-modules with $\{M_n\}_{n\in\mathbb{Z}}$ and $\{N_n\}_{n\in\mathbb{Z}}$ as their respective gradings; further, let

$$\lambda: M \to N \tag{8.2.2}$$

be a degree-preserving R-linear mapping, that is to say λ is a homomorphism of *graded modules*. By (8.1.2), λ^* maps N^* into M^* and it is easy to see that $\lambda^*(N_k^*) \subseteq M_k^*$ for all k. Accordingly there is induced a homomorphism

$$\lambda^\dagger: N^\dagger \to M^\dagger \tag{8.2.3}$$

of graded modules; moreover if $\lambda_k: M_k \to N_k$ is the homomorphism in degree k induced by λ, then the homomorphism $N_k^* \to M_k^*$ induced by $\lambda^\dagger$ is λ_k^*. Evidently

$$(\mathrm{id}_M)^\dagger = \mathrm{id}_{M^\dagger} \tag{8.2.4}$$

and if $\mu: N \to K$ is another homomorphism of graded modules, then

$$(\mu \circ \lambda)^\dagger = \lambda^\dagger \circ \mu^\dagger. \tag{8.2.5}$$

Thus the formation of graded duals is a contravariant functor from the category of graded R-modules (and their degree-preserving homomorphisms) to itself. Naturally if $\lambda: M \to N$ is an isomorphism of graded modules, then so too is $\lambda^\dagger: N^\dagger \to M^\dagger$ and

$$(\lambda^\dagger)^{-1} = (\lambda^{-1})^\dagger. \tag{8.2.6}$$

We next consider the homomorphism $M \to M^{**}$ of (8.1.7). This associates with each element of M a linear form on M^* and hence, by restriction, a linear form on $M^\dagger$. Thus we arrive at a homomorphism $M \to (M^\dagger)^*$ and now an easy verification shows that the image of M belongs to $(M^\dagger)^\dagger$. Accordingly we have a homomorphism

$$M \to M^{\dagger\dagger} \tag{8.2.7}$$

concerning which it should be noted that degrees are preserved, and that if we restrict ourselves to degree n we recover the familiar mapping $M_n \to M_n^{**}$ of M_n into its double dual. It follows that if $\lambda: M \to N$ is a homomorphism of graded modules, then

$$\begin{array}{ccc} M & \longrightarrow & M^{\dagger\dagger} \\ {\scriptstyle\lambda}\big\downarrow & & \big\downarrow{\scriptstyle\lambda^{\dagger\dagger}} \\ N & \longrightarrow & N^{\dagger\dagger} \end{array} \tag{8.2.8}$$

is a commutative diagram.

Lemma 3. *Let M be a graded R-module with finite free components. Then the homomorphism $M \to M^{\dagger\dagger}$ of* (8.2.7) *is an isomorphism of graded modules.*

This follows from Lemma 1 as soon as we consider components.

Now suppose that $M^{(1)}, M^{(2)}, \ldots, M^{(p)}$ are graded R-modules. We can turn $M^{(1)} \otimes M^{(2)} \otimes \cdots \otimes M^{(p)}$ and $M^{(1)\dagger} \otimes M^{(2)\dagger} \otimes \cdots \otimes M^{(p)\dagger}$ into graded modules by giving them the usual total gradings. Let $f_1 \in M^{(1)\dagger}$, $f_2 \in M^{(2)\dagger}$ and so on. Then $\mu_R^{(p)} \circ (f_1 \otimes f_2 \otimes \cdots \otimes f_p)$, where $\mu_R^{(p)}$ denotes p-fold multiplication, is a linear form on $M^{(1)} \otimes M^{(2)} \otimes \cdots \otimes M^{(p)}$ and among the submodules $M_{i_1}^{(1)} \otimes M_{i_2}^{(2)} \otimes \cdots \otimes M_{i_p}^{(p)}$, of $M^{(1)} \otimes \cdots \otimes M^{(p)}$, there are only finitely many on which $\mu_R^{(p)} \circ (f_1 \otimes f_2 \otimes \cdots \otimes f_p)$ does not vanish. Consequently the linear form belongs to $(M^{(1)} \otimes M^{(2)} \otimes \cdots \otimes M^{(p)})^\dagger$. It follows that there is an R-linear mapping

$$M^{(1)\dagger} \otimes M^{(2)\dagger} \otimes \cdots \otimes M^{(p)\dagger} \to (M^{(1)} \otimes M^{(2)} \otimes \cdots \otimes M^{(p)})^\dagger \tag{8.2.9}$$

in which $f_1 \otimes f_2 \otimes \cdots \otimes f_p$ (considered as an element of the left-hand side) is mapped into $\mu_R^{(p)} \circ (f_1 \otimes f_2 \otimes \cdots \otimes f_p)$. But $M^{(1)\dagger} \otimes M^{(2)\dagger} \otimes \cdots \otimes M^{(p)\dagger}$ is the direct sum of its submodules $M_{i_1}^{(1)*} \otimes M_{i_2}^{(2)*} \otimes \cdots \otimes M_{i_p}^{(p)*}$, and $(M_{i_1}^{(1)} \otimes M_{i_2}^{(2)} \otimes \cdots \otimes M_{i_p}^{(p)})^*$ is a submodule of the component of degree $i_1 + i_2 + \cdots + i_p$ of $(M^{(1)} \otimes M^{(2)} \otimes \cdots \otimes M^{(p)})^\dagger$. Furthermore (8.2.9) induces the homomorphism

$$M_{i_1}^{(1)*} \otimes M_{i_2}^{(2)*} \otimes \cdots \otimes M_{i_p}^{(p)*} \to (M_{i_1}^{(1)} \otimes M_{i_2}^{(2)} \otimes \cdots \otimes M_{i_p}^{(p)})^* \tag{8.2.10}$$

which operates exactly as in (8.1.11). In particular we see that (8.2.9) is a homomorphism of graded modules.

Lemma 4. *Let $M^{(1)}, M^{(2)}, \ldots, M^{(p)}$ be graded modules with finite free components, and let all the gradings be non-negative. Then the homomorphism*

$$M^{(1)\dagger} \otimes M^{(2)\dagger} \otimes \cdots \otimes M^{(p)\dagger} \to (M^{(1)} \otimes M^{(2)} \otimes \cdots \otimes M^{(p)})^\dagger$$

(*see* (8.2.9)) *is an isomorphism of graded modules.*

Proof. Since the components are finite free modules all the homomorphisms (8.2.10) are isomorphisms; and, because the gradings are non-negative, the component of $(M^{(1)} \otimes M^{(2)} \otimes \cdots \otimes M^{(p)})^\dagger$ of degree k is the direct *sum* of its submodules

$$(M_{i_1}^{(1)} \otimes M_{i_2}^{(2)} \otimes \cdots \otimes M_{i_p}^{(p)})^* \quad (i_1+i_2+\cdots+i_p=k)$$

(see (8.1.9)). The lemma follows.

We next consider some important situations involving homomorphisms like (8.2.9). First suppose that we have homomorphisms

$$\lambda_i: M^{(i)} \to N^{(i)} \quad (i=1,2,\ldots,p)$$

of graded modules. It is then easily verified that

$$\begin{array}{ccc} N^{(1)\dagger} \otimes N^{(2)\dagger} \otimes \cdots \otimes N^{(p)\dagger} & \longrightarrow & (N^{(1)} \otimes N^{(2)} \otimes \cdots \otimes N^{(p)})^\dagger \\ \Big\downarrow {\scriptstyle \lambda_1^\dagger \otimes \cdots \otimes \lambda_p^\dagger} & & \Big\downarrow {\scriptstyle (\lambda_1 \otimes \cdots \otimes \lambda_p)^\dagger} \\ M^{(1)\dagger} \otimes M^{(2)\dagger} \otimes \cdots \otimes M^{(p)\dagger} & \longrightarrow & (M^{(1)} \otimes M^{(2)} \otimes \cdots \otimes M^{(p)})^\dagger \end{array} \tag{8.2.11}$$

is a commutative diagram. Of course, it is understood that the horizontal mappings are derived from (8.2.9).

By Lemma 4, if all the modules have non-negative gradings and finite free components, then the horizontal mappings in (8.2.11) are isomorphisms. Hence in these circumstances we may make the identifications

$$N^{(1)\dagger} \otimes N^{(2)\dagger} \otimes \cdots \otimes N^{(p)\dagger} = (N^{(1)} \otimes N^{(2)} \otimes \cdots \otimes N^{(p)})^\dagger$$

and

$$M^{(1)\dagger} \otimes M^{(2)\dagger} \otimes \cdots \otimes M^{(p)\dagger} = (M^{(1)} \otimes M^{(2)} \otimes \cdots \otimes M^{(p)})^\dagger.$$

The commutative property of the diagram then ensures that

$$\lambda_1^\dagger \otimes \lambda_2^\dagger \otimes \cdots \otimes \lambda_p^\dagger = (\lambda_1 \otimes \lambda_2 \otimes \cdots \otimes \lambda_p)^\dagger. \tag{8.2.12}$$

Another situation which will concern us arises in the following way. Let $M^{(1)}, M^{(2)}, \ldots, M^{(p)}$ and $K^{(1)}, K^{(2)}, \ldots, K^{(q)}$ be graded R-modules. The homomorphisms

$$M^{(1)\dagger} \otimes M^{(2)\dagger} \otimes \cdots \otimes M^{(p)\dagger} \to (M^{(1)} \otimes M^{(2)} \otimes \cdots \otimes M^{(p)})^\dagger$$

and

$$K^{(1)\dagger} \otimes K^{(2)\dagger} \otimes \cdots \otimes K^{(q)\dagger} \to (K^{(1)} \otimes K^{(2)} \otimes \cdots \otimes K^{(q)})^\dagger$$

induce a homomorphism

$$\begin{aligned}(M^{(1)\dagger} \otimes \cdots \otimes M^{(p)\dagger}) \otimes (K^{(1)\dagger} \otimes \cdots \otimes K^{(q)\dagger}) \\ \to (M^{(1)} \otimes \cdots \otimes M^{(p)})^\dagger \otimes (K^{(1)} \otimes \cdots \otimes K^{(q)})^\dagger\end{aligned}$$

of graded modules. But, again from (8.2.9), we have a homomorphism

$$\begin{aligned}(M^{(1)} \otimes \cdots \otimes M^{(p)})^\dagger \otimes (K^{(1)} \otimes \cdots \otimes K^{(q)})^\dagger \\ \to ((M^{(1)} \otimes \cdots \otimes M^{(p)}) \otimes (K^{(1)} \otimes \cdots \otimes K^{(q)}))^\dagger\end{aligned}$$

and so we arrive at a mapping

$$(M^{(1)\dagger} \otimes \cdots \otimes M^{(p)\dagger}) \otimes (K^{(1)\dagger} \otimes \cdots \otimes K^{(q)\dagger})$$
$$\to ((M^{(1)} \otimes \cdots \otimes M^{(p)}) \otimes (K^{(1)} \otimes \cdots \otimes K^{(q)}))^{\dagger}. \tag{8.2.13}$$

This too is a homomorphism of graded modules.

Now consider the diagram

$$\begin{array}{ccc} M^{(1)\dagger} \otimes \cdots \otimes M^{(p)\dagger} \otimes K^{(1)\dagger} \otimes \cdots \otimes K^{(q)\dagger} & \longrightarrow & (M^{(1)} \otimes \cdots \otimes M^{(p)} \otimes K^{(1)} \otimes \cdots \otimes K^{(q)})^{\dagger} \\ \downarrow \approx & & \downarrow \approx \\ (M^{(1)\dagger} \otimes \cdots \otimes M^{(p)\dagger}) \otimes (K^{(1)\dagger} \otimes \cdots \otimes K^{(q)\dagger}) & \longrightarrow & ((M^{(1)} \otimes \cdots \otimes M^{(p)}) \otimes (K^{(1)} \otimes \cdots \otimes K^{(q)}))^{\dagger} \end{array} \tag{8.2.14}$$

(Here the upper horizontal mapping comes from (8.2.9) and the lower one from (8.2.13). The left vertical isomorphism comes from Theorem 1 of Chapter 2 whereas the one on the right-hand side is the graded dual of the isomorphism

$$(M^{(1)} \otimes \cdots \otimes M^{(p)}) \otimes (K^{(1)} \otimes \cdots \otimes K^{(q)})$$
$$\xrightarrow{\sim} M^{(1)} \otimes \cdots \otimes M^{(p)} \otimes K^{(1)} \otimes \cdots \otimes K^{(q)}$$

provided by the same result.) It is a straightforward matter to check that (8.2.14) is commutative. We shall describe this fact by saying that *the taking of graded duals is compatible with the associative property of tensor products* as embodied in Theorem 1 of Chapter 2. Note that all the mappings preserve degrees and that when the gradings are non-negative and all components are finite free modules, the horizontal homomorphisms are actually isomorphisms.

There is one more result connected with (8.2.9) that it is useful to place on record. Let $M^{(1)}, M^{(2)}, \ldots, M^{(p)}$ be graded modules and let $\Lambda(M^{(1)}, M^{(2)}, \ldots, M^{(p)})$ denote the isomorphism

$$(M^{(1)} \otimes \cdots \otimes M^{(p)}) \otimes (M^{(1)} \otimes \cdots \otimes M^{(p)})$$
$$\xrightarrow{\sim} (M^{(1)} \otimes M^{(1)}) \otimes \cdots \otimes (M^{(p)} \otimes M^{(p)}) \tag{8.2.15}$$

of graded modules in which

$$(m_{i_1} \otimes m_{i_2} \otimes \cdots \otimes m_{i_p}) \otimes (m'_{j_1} \otimes m'_{j_2} \otimes \cdots \otimes m'_{j_p})$$

is mapped into $(m_{i_1} \otimes m'_{j_1}) \otimes (m_{i_2} \otimes m'_{j_2}) \otimes \cdots \otimes (m_{i_p} \otimes m'_{j_p})$. (The notation is the normal one, i.e. m_{i_1} and m'_{j_1} belong to $M^{(1)}_{i_1}$ and $M^{(1)}_{j_1}$ respectively and so on.) The isomorphism

$$(M^{(1)} \otimes M^{(1)}) \otimes \cdots \otimes (M^{(p)} \otimes M^{(p)})$$
$$\xrightarrow{\sim} (M^{(1)} \otimes \cdots \otimes M^{(p)}) \otimes (M^{(1)} \otimes \cdots \otimes M^{(p)}) \tag{8.2.16}$$

that is inverse to (8.2.15) will be denoted by $V(M^{(1)}, M^{(2)}, \ldots, M^{(p)})$.

It is now a simple matter to verify that the diagram

$$\begin{array}{ccc}
(M^{(1)\dagger} \otimes M^{(1)\dagger}) \otimes \cdots \otimes (M^{(p)\dagger} \otimes M^{(p)\dagger}) & \longrightarrow & ((M^{(1)} \otimes M^{(1)}) \otimes \cdots \otimes (M^{(p)} \otimes M^{(p)}))^{\dagger} \\
\downarrow V(M^{(1)\dagger}, \ldots, M^{(p)\dagger}) & & \downarrow \Lambda(M^{(1)}, \ldots, M^{(p)})^{\dagger} \\
(M^{(1)\dagger} \otimes \cdots \otimes M^{(p)\dagger}) \otimes (M^{(1)\dagger} \otimes \cdots \otimes M^{(p)\dagger}) & \longrightarrow & ((M^{(1)} \otimes \cdots \otimes M^{(p)}) \otimes (M^{(1)} \otimes \cdots \otimes M^{(p)}))^{\dagger}
\end{array} \tag{8.2.17}$$

is commutative. (Of course, the horizontal mappings are obtained by using homomorphisms of the kind typified by (8.2.9).) Once again, all the mappings preserve degrees.

Should it happen that $M^{(1)}, M^{(2)}, \ldots, M^{(p)}$ have non-negative gradings and finite free components, then the horizontal mappings in (8.2.17) will be isomorphisms so that in each case we can identify the domain with the codomain. If this is done we find that

$$V(M^{(1)\dagger}, M^{(2)\dagger}, \ldots, M^{(p)\dagger}) = \Lambda(M^{(1)}, M^{(2)}, \ldots, M^{(p)})^{\dagger} \tag{8.2.18}$$

and now, by taking inverses, we can add the relation

$$\Lambda(M^{(1)\dagger}, M^{(2)\dagger}, \ldots, M^{(p)\dagger}) = V(M^{(1)}, M^{(2)}, \ldots, M^{(p)})^{\dagger}. \tag{8.2.19}$$

The isomorphisms (8.2.15) and (8.2.16) may be modified by introducing signs just as we did in (7.1.3) and (7.5.6). Let $\underline{\Lambda}(M^{(1)}, \ldots, M^{(p)})$ and $\underline{V}(M^{(1)}, \ldots, M^{(p)})$ be the results of making these changes. We then find that when $M^{(1)}$, $M^{(2)}$, $\ldots$, $M^{(p)}$ have non-negative gradings and finite free components we can replace (8.2.18) and (8.2.19) by

$$\underline{V}(M^{(1)\dagger}, M^{(2)\dagger}, \ldots, M^{(p)\dagger}) = \underline{\Lambda}(M^{(1)}, M^{(2)}, \ldots, M^{(p)})^{\dagger} \tag{8.2.20}$$

and

$$\underline{\Lambda}(M^{(1)\dagger}, M^{(2)\dagger}, \ldots, M^{(p)\dagger}) = \underline{V}(M^{(1)}, M^{(2)}, \ldots, M^{(p)})^{\dagger} \tag{8.2.21}$$

respectively.

Our final comments in this section involve much simpler situations. Evidently if the R-module K is graded trivially, then $K^{\dagger}$ is just K^* with the trivial grading. In particular

$$R^{\dagger} = R^* \approx R, \tag{8.2.22}$$

where the isomorphism $R^* \approx R$ comes from (8.1.6). We shall use (8.2.22) to identify $R^{\dagger}$ and R as graded modules.

Now consider the homomorphisms

$$\mu_R^{(p)}: R \otimes R \otimes \cdots \otimes R \to R,$$
$$\Delta_R^{(p)}: R \to R \otimes R \otimes \cdots \otimes R,$$

as defined in (7.4.5) and (7.1.4) respectively, where R is now regarded as a trivially graded module. Then

$$\mu_R^{(p)\dagger}: R^\dagger \to (R \otimes R \otimes \cdots \otimes R)^\dagger = R^\dagger \otimes R^\dagger \otimes \cdots \otimes R^\dagger$$

and

$$\Delta_R^{(p)\dagger}: R^\dagger \otimes R^\dagger \otimes \cdots \otimes R^\dagger = (R \otimes R \otimes \cdots \otimes R)^\dagger \to R^\dagger,$$

so that the identification $R^\dagger = R$ turns $\mu_R^{(p)\dagger}$ into a mapping $R \to R \otimes R \otimes \cdots \otimes R$ and it turns $\Delta_R^{(p)\dagger}$ into a similar mapping in the reverse direction. On this understanding we find that

$$\mu_R^{(p)\dagger} = \Delta_R^{(p)} \tag{8.2.23}$$

and

$$\Delta_R^{(p)\dagger} = \mu_R^{(p)}. \tag{8.2.24}$$

8.3 Graded duals of algebras and coalgebras

In Section (7.7) we considered an algebra that was formed by linear forms on a coalgebra, and now that we are about to discuss the graded duals of algebras and coalgebras, it is convenient to point out the connections between the present discussion and the ideas that were developed earlier. It is for this reason that we begin with coalgebras.

Suppose then that (A, Δ, ε) is a non-negatively graded coalgebra whose components are finite free modules. Then $\Delta^\dagger: (A \otimes A)^\dagger \to A^\dagger$ and $\varepsilon^\dagger: R^\dagger \to A^\dagger$. But, by Lemma 4, $(A \otimes A)^\dagger$ can be identified with $A^\dagger \otimes A^\dagger$ and, by (8.2.22), $R^\dagger$ can be identified with R. In this way we arrive at degree-preserving R-linear mappings

$$\Delta^\dagger: A^\dagger \otimes A^\dagger \to A^\dagger \tag{8.3.1}$$

and

$$\varepsilon^\dagger: R \to A^\dagger. \tag{8.3.2}$$

Theorem 1. *Let (A, Δ, ε) be a graded R-coalgebra with a non-negative grading and finite free components. Then $(A^\dagger, \Delta^\dagger, \varepsilon^\dagger)$ is a graded R-algebra.*

Proof. Because $(A^\dagger, \Delta^\dagger, \varepsilon^\dagger)$ will turn out to be a special case of a graded algebra that was encountered in the last chapter, we shall use an argument that exhibits this relationship rather than one which is based directly on the definition.

Let $f, g \in A^\dagger$. By (8.2.9), when $A^\dagger \otimes A^\dagger$ is identified with $(A \otimes A)^\dagger$ the element $f \otimes g$ of the former becomes $\mu_R \circ (f \otimes g)$, where μ_R denotes the multiplication mapping of R. Thus for $\Delta^\dagger: A^\dagger \otimes A^\dagger \to A^\dagger$ we have

$$\Delta^\dagger(f \otimes g) = \mu_R \circ (f \otimes g) \circ \Delta. \tag{8.3.3}$$

On the other hand

$$\varepsilon^\dagger(1) = \mathrm{id}_R \circ \varepsilon = \varepsilon. \tag{8.3.4}$$

But already, in Theorem 8 of Chapter 7, $A^\dagger$ has been shown to have a natural structure as a graded R-algebra and now (8.3.3) and (8.3.4) show that $\Delta^\dagger$ and $\varepsilon^\dagger$ are the multiplication and unit mappings of the previously encountered algebra. This proves the theorem. It should be observed, however, that Theorem 8 of Chapter 7 was proved under much more general conditions.

Definition. *With the above notation the graded algebra* $(A^\dagger, \Delta^\dagger, \varepsilon^\dagger)$ *is called the 'graded dual' of the graded coalgebra* (A, Δ, ε).

Note that the graded dual also has a non-negative grading and finite free components.

Theorem 2. *Let A and B be graded coalgebras with non-negative gradings and finite free components. Further, let $\lambda: A \to B$ be a homomorphism of graded coalgebras. Then $\lambda^\dagger: B^\dagger \to A^\dagger$ is a homomorphism of graded algebras.*

This follows at once from Exercise 6 of Chapter 7. However, a direct proof can be obtained by adapting the arguments, given below, to establish the dual result (Theorem 4).

We turn now to algebras. Suppose that (A, μ, η) is a non-negatively graded algebra with finite free components. Then $\mu^\dagger$ maps $A^\dagger$ into $(A \otimes A)^\dagger$ and $\eta^\dagger$ maps $A^\dagger$ into $R^\dagger$. Consequently if we make the identifications $(A \otimes A)^\dagger = A^\dagger \otimes A^\dagger$ and $R^\dagger = R$, then we obtain R-linear mappings

$$\mu^\dagger: A^\dagger \to A^\dagger \otimes A^\dagger \tag{8.3.5}$$

and

$$\eta^\dagger: A^\dagger \to R \tag{8.3.6}$$

which preserve the degrees of homogeneous elements.

Theorem 3. *Let (A, μ, η) be a graded R-algebra, where the grading is non-negative and has finite free components. Then $(A^\dagger, \mu^\dagger, \eta^\dagger)$ is a graded R-coalgebra.*

Proof. Since $\mu \circ (\mu \otimes A) = \mu \circ (A \otimes \mu)$ we have

$$(\mu \circ (\mu \otimes A))^\dagger = (\mu \circ (A \otimes \mu))^\dagger. \tag{8.3.7}$$

Now, to be quite precise, $\mu \circ (\mu \otimes A)$ is the result of combining the homomorphisms

$$A \otimes A \otimes A \xrightarrow{\sim} (A \otimes A) \otimes A \xrightarrow{\mu \otimes A} A \otimes A \xrightarrow{\mu} A$$

so that $(\mu \circ (\mu \otimes A))^\dagger$ is the total mapping

$$A^\dagger \xrightarrow{\mu^\dagger} (A \otimes A)^\dagger \xrightarrow{(\mu \otimes A)^\dagger} ((A \otimes A) \otimes A)^\dagger$$

$$\xrightarrow{\sim} (A \otimes A \otimes A)^\dagger. \tag{8.3.8}$$

Let us make the identifications $(A \otimes A)^{\dagger} = A^{\dagger} \otimes A^{\dagger}$ and

$$((A \otimes A) \otimes A)^{\dagger} = (A \otimes A)^{\dagger} \otimes A^{\dagger} = (A^{\dagger} \otimes A^{\dagger}) \otimes A^{\dagger}.$$

Then the first mapping in (8.3.8) becomes the mapping (8.3.5) and, by (8.2.12), the second one is turned into $\mu^{\dagger} \otimes A^{\dagger}$. Accordingly (8.3.8) becomes

$$\begin{array}{ccccc} A^{\dagger} & \xrightarrow{\mu^{\dagger}} & A^{\dagger} \otimes A^{\dagger} & \xrightarrow{\mu^{\dagger} \otimes A^{\dagger}} & (A^{\dagger} \otimes A^{\dagger}) \otimes A^{\dagger} \\ & & & & \| \\ & & & & ((A \otimes A) \otimes A)^{\dagger} \xrightarrow{\sim} (A \otimes A \otimes A)^{\dagger}; \end{array}$$

and if next we identify $(A \otimes A \otimes A)^{\dagger}$ with $A^{\dagger} \otimes A^{\dagger} \otimes A^{\dagger}$, then the final mapping is the module-isomorphism $(A^{\dagger} \otimes A^{\dagger}) \otimes A^{\dagger} \xrightarrow{\sim} A^{\dagger} \otimes A^{\dagger} \otimes A^{\dagger}$ (see (8.2.14)). Thus $(\mu \circ (\mu \otimes A))^{\dagger}$, considered as a mapping of $A^{\dagger}$ into $A^{\dagger} \otimes A^{\dagger} \otimes A^{\dagger}$, is simply $(\mu^{\dagger} \otimes A^{\dagger}) \circ \mu^{\dagger}$; likewise

$$(\mu \circ (A \otimes \mu))^{\dagger} \colon A^{\dagger} \to A^{\dagger} \otimes A^{\dagger} \otimes A^{\dagger}$$

is none other than $(A^{\dagger} \otimes \mu^{\dagger}) \circ \mu^{\dagger}$. Accordingly

$$(\mu^{\dagger} \otimes A^{\dagger}) \circ \mu^{\dagger} = (A^{\dagger} \otimes \mu^{\dagger}) \circ \mu^{\dagger}$$

so that $\mu^{\dagger} \colon A^{\dagger} \to A^{\dagger} \otimes A^{\dagger}$ is coassociative.

Next, the total mapping

$$A \xrightarrow{\sim} R \otimes A \xrightarrow{\eta \otimes A} A \otimes A \xrightarrow{\mu} A$$

is just the identity mapping of A and therefore

$$A^{\dagger} \xrightarrow{\mu^{\dagger}} (A \otimes A)^{\dagger} \xrightarrow{(\eta \otimes A)^{\dagger}} (R \otimes A)^{\dagger} \xrightarrow{\sim} A^{\dagger} \qquad (8.3.9)$$

is the identity mapping of $A^{\dagger}$. If now we put $(A \otimes A)^{\dagger} = A^{\dagger} \otimes A^{\dagger}$ and

$$(R \otimes A)^{\dagger} = R^{\dagger} \otimes A^{\dagger} = R \otimes A^{\dagger}$$

we find that (8.3.9) becomes

$$\begin{array}{ccccc} A^{\dagger} & \xrightarrow{\mu^{\dagger}} & A^{\dagger} \otimes A^{\dagger} & \xrightarrow{\eta^{\dagger} \otimes A^{\dagger}} & R \otimes A^{\dagger} \\ & & & & \| \\ & & & & (R \otimes A)^{\dagger} \xrightarrow{\sim} A^{\dagger}. \end{array}$$

Consider the final mapping in this sequence. If $f \in A^{\dagger}$, then $1 \otimes f$ (regarded as a member of $(R \otimes A)^{\dagger}$) maps $r \otimes a$ into $rf(a)$, hence the image of $1 \otimes f$ in $A^{\dagger}$ is just f itself. Accordingly the total mapping

$$A^{\dagger} \xrightarrow{\mu^{\dagger}} A^{\dagger} \otimes A^{\dagger} \xrightarrow{\eta^{\dagger} \otimes A^{\dagger}} R \otimes A^{\dagger} \xrightarrow{\sim} A^{\dagger}$$

is the identity mapping of $A^{\dagger}$. Thus $(\eta^{\dagger} \otimes A^{\dagger}) \circ \mu^{\dagger}$ is the canonical isomorphism $A^{\dagger} \xrightarrow{\sim} R \otimes A^{\dagger}$ and similar considerations show that $(A^{\dagger} \otimes \eta^{\dagger}) \circ \mu^{\dagger}$ is the isomorphism $A^{\dagger} \xrightarrow{\sim} A^{\dagger} \otimes R$. The proof is now complete.

Definition. *The graded coalgebra* $(A^\dagger, \mu^\dagger, \eta^\dagger)$ *of Theorem 3 is called the 'graded dual' of the graded algebra* (A, μ, η).

Naturally the grading on the graded dual is non-negative and the components are finite free modules.

Theorem 4. *Let* A *and* B *be graded* R*-algebras with non-negative gradings and finite free components, and let* $\lambda: A \to B$ *be a homomorphism of graded algebras. Then* $\lambda^\dagger: B^\dagger \to A^\dagger$ *is a homomorphism of graded coalgebras.*

Proof. If we apply graded duality to the commutative diagram

$$\begin{array}{ccc} A \otimes A & \xrightarrow{\mu_A} & A \\ {\scriptstyle \lambda\otimes\lambda}\downarrow & & \downarrow{\scriptstyle \lambda} \\ B \otimes B & \xrightarrow{\mu_B} & B \end{array}$$

the result is a new commutative diagram; and if we then make the identifications $(A \otimes A)^\dagger = A^\dagger \otimes A^\dagger$ and $(B \otimes B)^\dagger = B^\dagger \otimes B^\dagger$, the new diagram becomes

$$\begin{array}{ccc} A^\dagger \otimes A^\dagger & \longleftarrow & A^\dagger \\ {\scriptstyle \lambda^\dagger\otimes\lambda^\dagger}\uparrow & & \uparrow{\scriptstyle \lambda^\dagger} \\ B^\dagger \otimes B^\dagger & \longleftarrow & B^\dagger \end{array}$$

where the horizontal mappings are comultiplications. Thus $\lambda^\dagger$ is compatible with comultiplication. A similar exercise carried out on the commutative diagram

$$\begin{array}{ccc} & \overset{\eta_A}{\nearrow} & A \\ R & & \downarrow{\scriptstyle \lambda} \\ & \underset{\eta_B}{\searrow} & B \end{array}$$

shows that $\lambda^\dagger$ also preserves counits. The proof is therefore complete.

We know that R is both a graded R-algebra and a graded R-coalgebra; indeed the results of this section are applicable to it. Consequently $R^\dagger$ is also a graded algebra and a graded coalgebra.

Lemma 5. *The isomorphism* $R^\dagger \approx R$ *of* (8.2.22) *is an isomorphism of graded algebras and graded coalgebras.*

The verification is trivial if we use (8.2.23) and (8.2.24).

We end this section by confirming that when graded duality is applied twice we return (essentially) to where we started.

Theorem 5. *Let A be a graded algebra, respectively coalgebra, with a non-negative grading and finite free components. Then the canonical isomorphism $A \xrightarrow{\sim} A^{\dagger\dagger}$ of graded modules (see Lemma 3) is actually an isomorphism of graded algebras, respectively coalgebras.*

Proof. We shall deal only with the case where A is a coalgebra; the other case can be dealt with similarly.

Let Δ and ε be the comultiplication and counit mappings of A. By (8.2.8), the diagram

$$\begin{array}{ccc} A & \longrightarrow & A^{\dagger\dagger} \\ {\scriptstyle\Delta}\big\downarrow & & \big\downarrow{\scriptstyle\Delta^{\dagger\dagger}} \\ A \otimes A & \longrightarrow & (A \otimes A)^{\dagger\dagger} \end{array}$$

is commutative. If we make the identifications $(A \otimes A)^{\dagger\dagger} = (A^\dagger \otimes A^\dagger)^\dagger = A^{\dagger\dagger} \otimes A^{\dagger\dagger}$ we find that $A \otimes A \rightarrow (A \otimes A)^{\dagger\dagger}$, considered as a homomorphism of $A \otimes A$ into $A^{\dagger\dagger} \otimes A^{\dagger\dagger}$, is the tensor product of the homomorphism $A \rightarrow A^{\dagger\dagger}$ with itself. This shows that $A \rightarrow A^{\dagger\dagger}$ is compatible with comultiplication.

Next we consider the commutative diagram

$$\begin{array}{ccc} A & \longrightarrow & A^{\dagger\dagger} \\ {\scriptstyle\varepsilon}\big\downarrow & & \big\downarrow{\scriptstyle\varepsilon^{\dagger\dagger}} \\ R & \longrightarrow & R^{\dagger\dagger} \end{array}$$

and observe that when we put

$$R^{\dagger\dagger} = R^\dagger = R$$

the mapping $R \rightarrow R^{\dagger\dagger}$ becomes the identity mapping of R. Accordingly $A \rightarrow A^{\dagger\dagger}$ preserves counits and with this the proof is complete.

8.4 Graded duals of Hopf algebras

We can extend the results of the last section to Hopf algebras, but first of all it is necessary to see how tensor products of graded algebras and coalgebras are effected by graded duality.

We begin with algebras. Let $A^{(1)}, A^{(2)}, \ldots, A^{(p)}$ be algebras with non-negative gradings whose components are finite free modules. Then $A^{(1)} \otimes A^{(2)} \otimes \cdots \otimes A^{(p)}$ is a graded algebra of the same kind. We propose to compare $(A^{(1)} \otimes A^{(2)} \otimes \cdots \otimes A^{(p)})^\dagger$ and $A^{(1)\dagger} \otimes A^{(2)\dagger} \otimes \cdots \otimes A^{(p)\dagger}$ as graded *coalgebras.*

Let $\Lambda(A^{(1)}, A^{(2)}, \ldots, A^{(p)})$ be the isomorphism

$$(A^{(1)} \otimes \cdots \otimes A^{(p)}) \otimes (A^{(1)} \otimes \cdots \otimes A^{(p)})$$
$$\xrightarrow{\sim} (A^{(1)} \otimes A^{(1)}) \otimes \cdots \otimes (A^{(p)} \otimes A^{(p)}) \qquad (8.4.1)$$

of graded modules which operates as in (8.2.15) and let the inverse isomorphism

$$(A^{(1)} \otimes A^{(1)}) \otimes \cdots \otimes (A^{(p)} \otimes A^{(p)})$$
$$\xrightarrow{\sim} (A^{(1)} \otimes \cdots \otimes A^{(p)}) \otimes (A^{(1)} \otimes \cdots \otimes A^{(p)}) \qquad (8.4.2)$$

be denoted by $V(A^{(1)}, A^{(2)}, \ldots, A^{(p)})$ (see (8.2.16)). If now μ_i and η_i are the multiplication and unit mappings of $A^{(i)}$, then the multiplication mapping of $A^{(1)} \otimes A^{(2)} \otimes \cdots \otimes A^{(p)}$ is

$$(\mu_1 \otimes \mu_2 \otimes \cdots \otimes \mu_p) \circ \Lambda(A^{(1)}, \ldots, A^{(p)})$$

and its unit mapping is

$$(\eta_1 \otimes \eta_2 \otimes \cdots \otimes \eta_p) \circ \Delta_R^{(p)}.$$

At this point let us make the identifications

$$(A^{(1)} \otimes A^{(2)} \otimes \cdots \otimes A^{(p)})^\dagger = A^{(1)\dagger} \otimes A^{(2)\dagger} \otimes \cdots \otimes A^{(p)\dagger},$$
$$((A^{(1)} \otimes A^{(1)}) \otimes \cdots \otimes (A^{(p)} \otimes A^{(p)}))^\dagger$$
$$= (A^{(1)\dagger} \otimes A^{(1)\dagger}) \otimes \cdots \otimes (A^{(p)\dagger} \otimes A^{(p)\dagger})$$

and

$$((A^{(1)} \otimes \cdots \otimes A^{(p)}) \otimes (A^{(1)} \otimes \cdots \otimes A^{(p)}))^\dagger$$
$$= (A^{(1)\dagger} \otimes \cdots \otimes A^{(p)\dagger}) \otimes (A^{(1)\dagger} \otimes \cdots \otimes A^{(p)\dagger}).$$

Then, using (8.2.18), we see that the comultiplication mapping of $(A^{(1)} \otimes A^{(2)} \otimes \cdots \otimes A^{(p)})^\dagger$ is

$$\Lambda(A^{(1)}, A^{(2)}, \ldots, A^{(p)})^\dagger \circ (\mu_1 \otimes \mu_2 \otimes \cdots \otimes \mu_p)^\dagger$$
$$= V(A^{(1)\dagger}, A^{(2)\dagger}, \ldots, A^{(p)\dagger}) \circ (\mu_1^\dagger \otimes \mu_2^\dagger \otimes \cdots \otimes \mu_p^\dagger) \qquad (8.4.3)$$

and that its counit mapping is

$$\Delta_R^{(p)\dagger} \circ (\eta_1 \otimes \eta_2 \otimes \cdots \otimes \eta_p)^\dagger$$
$$= \mu_R^{(p)} \circ (\eta_1^\dagger \otimes \eta_2^\dagger \otimes \cdots \otimes \eta_p^\dagger) \qquad (8.4.4)$$

(see (8.2.24)).

Now the right-hand sides of (8.4.3) and (8.4.4) are none other than the comultiplication and counit mappings of $A^{(1)\dagger} \otimes \cdots \otimes A^{(p)\dagger}$. Consequently we have proved that the natural isomorphism

$$A^{(1)\dagger} \otimes A^{(2)\dagger} \otimes \cdots \otimes A^{(p)\dagger}$$
$$\xrightarrow{\sim} (A^{(1)} \otimes A^{(2)} \otimes \cdots \otimes A^{(p)})^\dagger \qquad (8.4.5)$$

of graded modules (see Lemma 4) is actually an isomorphism of graded coalgebras.

Of course (8.4.5) may also be regarded as a module-isomorphism

$$A^{(1)\dagger} \underline{\otimes} A^{(2)\dagger} \underline{\otimes} \cdots \underline{\otimes} A^{(p)\dagger}$$

$$\xrightarrow{\sim} (A^{(1)} \underline{\otimes} A^{(2)} \underline{\otimes} \cdots \underline{\otimes} A^{(p)})^{\dagger} \tag{8.4.6}$$

and this too can be shown, in very much the same way, to be an isomorphism of graded coalgebras. We record both these results in

Theorem 6. *Let $A^{(1)}, A^{(2)}, \ldots, A^{(p)}$ be non-negatively graded algebras whose components are finite free modules. Then the graded coalgebras $A^{(1)\dagger} \otimes A^{(2)\dagger} \otimes \cdots \otimes A^{(p)\dagger}$ and $(A^{(1)} \otimes A^{(2)} \otimes \cdots \otimes A^{(p)})^{\dagger}$ can be identified by means of* (8.4.5); *and the graded coalgebras $A^{(1)\dagger} \underline{\otimes} A^{(2)\dagger} \underline{\otimes} \cdots \underline{\otimes} A^{(p)\dagger}$ and $(A^{(1)} \underline{\otimes} A^{(2)} \underline{\otimes} \cdots \underline{\otimes} A^{(p)})^{\dagger}$ can also be identified by means of the same mapping* (*see* (8.4.6)).

Naturally this result has an analogue in which the roles of algebras and coalgebras are interchanged. This is recorded as Theorem 7. The adjustments that need to be made to the proof of Theorem 6 are perfectly straightforward so no details will be given.

Theorem 7. *Let $A^{(1)}, A^{(2)}, \ldots, A^{(p)}$ be non-negatively graded coalgebras with finite free components. Then the isomorphism*

$$A^{(1)\dagger} \otimes A^{(2)\dagger} \otimes \cdots \otimes A^{(p)\dagger} \approx (A^{(1)} \otimes A^{(2)} \otimes \cdots \otimes A^{(p)})^{\dagger},$$

(*of graded modules*) *provided by Lemma 4 enables us to make the following identifications of graded algebras:*

(i) $A^{(1)\dagger} \otimes A^{(2)\dagger} \otimes \cdots \otimes A^{(p)\dagger}$ *with* $(A^{(1)} \otimes A^{(2)} \otimes \cdots \otimes A^{(p)})^{\dagger}$

and

(ii) $A^{(1)\dagger} \underline{\otimes} A^{(2)\dagger} \underline{\otimes} \cdots \underline{\otimes} A^{(p)\dagger}$ *with* $(A^{(1)} \underline{\otimes} A^{(2)} \underline{\otimes} \cdots \underline{\otimes} A^{(p)})^{\dagger}$.

We are now in a position to discuss the graded duals of Hopf algebras. By Theorems 1 and 3, if $(\eta, \mu, A, \Delta, \varepsilon)$ is a non-negatively graded Hopf algebra with finite free components, then $(A^{\dagger}, \Delta^{\dagger}, \varepsilon^{\dagger})$ is a graded algebra and $(A^{\dagger}, \mu^{\dagger}, \eta^{\dagger})$ is a graded coalgebra. Next, because A is a Hopf algebra, $\mu: A \otimes A \to A$ and $\eta: R \to A$ are homomorphisms of graded coalgebras and so (Theorem 2) $\mu^{\dagger}: A^{\dagger} \to (A \otimes A)^{\dagger}$ and $\eta^{\dagger}: A^{\dagger} \to R^{\dagger}$ are homomorphisms of graded algebras. But Theorem 6 and Lemma 5 show that the familiar module-isomorphisms $(A \otimes A)^{\dagger} \approx A^{\dagger} \otimes A^{\dagger}$ and $R^{\dagger} \approx R$ are, in fact, isomorphisms of graded algebras. Consequently $\mu^{\dagger}: A^{\dagger} \to A^{\dagger} \otimes A^{\dagger}$ and $\eta^{\dagger}: A^{\dagger} \to R$ are algebra-homomorphisms.

Theorem 8. *Let $(\eta, \mu, A, \Delta, \varepsilon)$ be a non-negatively graded Hopf algebra with finite free components. Then $(\varepsilon^{\dagger}, \Delta^{\dagger}, A^{\dagger}, \mu^{\dagger}, \eta^{\dagger})$ is a graded Hopf algebra of the same kind.*

This follows immediately from our previous remarks as soon as we invoke Lemma 5 Cor. of Chapter 7. Naturally we refer to $(\varepsilon^\dagger, \Delta^\dagger, A^\dagger, \mu^\dagger, \eta^\dagger)$ as the *graded dual* of the Hopf algebra $(\eta, \mu, A, \Delta, \varepsilon)$.

Finally, by using Lemma 6 Cor. of Chapter 7, we can adapt the above argument so that it applies to *modified* Hopf algebras. We thus obtain

Theorem 9. *Let $(\eta, \mu, A, \Delta, \varepsilon)$ be a modified Hopf algebra with a non-negative grading and finite free components. Then $(\varepsilon^\dagger, \Delta^\dagger, A^\dagger, \mu^\dagger, \eta^\dagger)$ is a modified Hopf algebra of the same kind.*

Finally let us see how R itself fits into this theory. It is, of course, both a graded Hopf algebra and a modified Hopf algebra, and therefore the same is true of $R^\dagger$. Furthermore, by Lemma 5, the usual module-isomorphism $R^\dagger \approx R$ is an isomorphism of Hopf algebras. Hence, under graded duality, the Hopf algebra R remains essentially unchanged.

8.5 Comments and exercises

Let $M_1, M_2, \ldots, M_p$ be R-modules. Then, by (8.1.11), we have a homomorphism

$$M_1^* \otimes M_2^* \otimes \cdots \otimes M_p^* \to (M_1 \otimes M_2 \otimes \cdots \otimes M_p)^*;$$

and if we take $M_1, M_2, \ldots, M_p$ to be the same module M, the homomorphism takes the form

$$T_p(M^*) \to T_p(M)^*, \tag{8.5.1}$$

where $T_p(M)$ denotes the p-th tensor power of M. In this section we shall consider certain natural homomorphisms $E_p(M^*) \to E_p(M)^*$ and $S_p(M^*) \to S_p(M)^*$.

By (7.11.8), if $G(M)$ is the Grassmann algebra of M, then M^* is the submodule of $G(M)$ formed by the homogeneous elements of degree one; and, by Theorem 13 of Chapter 7, $G(M)$ is an anticommutative algebra. It follows that the inclusion mapping $M^* \to G(M)$ extends to a homomorphism

$$E(M^*) \to G(M) \tag{8.5.2}$$

of R-algebras which preserves degrees. In degree zero (8.5.2) induces a homomorphism $R \to \mathrm{Hom}(R, R)$ which maps the identity element of R into the identity mapping of R; and in degree $p \geq 1$ it induces an R-linear mapping

$$E_p(M^*) \to E_p(M)^*. \tag{8.5.3}$$

This is one of the homomorphisms we are seeking. Let us see how it operates.

Suppose that $f_1, f_2, \ldots, f_p$ belong to M^* and let the image of $f_1 \wedge f_2 \wedge \cdots \wedge f_p$ under (8.5.3) be ϕ. Since (8.5.2) is a homomorphism of algebras, ϕ is the product of $f_1, f_2, \ldots, f_p$ in $G(M)$. Consequently, if $m_1, m_2, \ldots, m_p$ belong to M, then (Chapter 7, Lemma 11)

$$\phi(m_1 \wedge m_2 \wedge \cdots \wedge m_p) = \begin{vmatrix} f_1(m_1) & f_1(m_2) & \cdots & f_1(m_p) \\ f_2(m_1) & f_2(m_2) & \cdots & f_2(m_p) \\ \vdots & \vdots & \vdots\vdots\vdots & \vdots \\ f_p(m_1) & f_p(m_2) & \cdots & f_p(m_p) \end{vmatrix}. \tag{8.5.4}$$

Thus the image of $f_1 \wedge f_2 \wedge \cdots \wedge f_p$ under (8.5.3) is the linear form on $E_p(M)$ that maps $m_1 \wedge m_2 \wedge \cdots \wedge m_p$ into the determinant of the matrix $\| f_i(m_j) \|$.

Now suppose that M is a finite free R-module with $b_1, b_2, \ldots, b_s$ as a base, and let the base of M^* that is dual to this be $f_1, f_2, \ldots, f_s$. Suppose that $1 \leq p \leq s$ and let $I = (i_1, i_2, \ldots, i_p)$ and $J = (j_1, j_2, \ldots, j_p)$ be strictly increasing sequences of integers between 1 and s. Finally put $b_I = b_{i_1} \wedge b_{i_2} \wedge \cdots \wedge b_{i_p}$ and $f_J = f_{j_1} \wedge f_{j_2} \wedge \cdots \wedge f_{j_p}$. Then the b_I form a base for $E_p(M)$ and the f_J a base for $E_p(M^*)$.

Exercise 1*. *Let the notation be as explained. Show that the base* $\{f_J\}$ *of* $E_p(M^*)$ *is mapped by* (8.5.3) *into the base of* $E_p(M)^*$ *that is dual to the base* $\{b_I\}$ *of* $E_p(M)$.

As usual, the asterisk attached to the number of the exercise indicates that a solution is provided in the next section.

It follows from Exercise 1 that *if* M *is a finite free module, then the homomorphism* $E_p(M^*) \to E_p(M)^*$ *of* (8.5.3) *is an isomorphism of modules, and that* $E(M^*) \to G(M)$ *(see* (8.5.2)) *is an isomorphism of graded algebras.*

In the case where M is a finite free module, $E(M)$ is not only a modified Hopf algebra with a non-negative grading, but also it has finite free components. Accordingly (Theorem 9) its graded dual $E(M)^\dagger$ is a modified Hopf algebra as well. Furthermore, the proof of Theorem 1 shows that the *algebras* $E(M)^\dagger$ and $G(M)$ coincide, so that by the last paragraph we have an isomorphism $E(M^*) \xrightarrow{\sim} E(M)^\dagger$ of graded algebras. However, more than this is true as the next exercise shows.

Exercise 2*. *If* M *is a finite free* R*-module, show that the mapping* $E(M^*) \to E(M)^\dagger$, *defined above, is an isomorphism of modified Hopf algebras.*

As is to be expected there are some analogous results connected with symmetric algebras though, as we shall see later, there is one important difference. We examine the details in the following paragraphs.

If M is an arbitrary R-module, then $S(M)$ is a graded Hopf algebra and so *a fortiori* it is a graded coalgebra; and we have already used Theorem 8 of Chapter 7 to derive from it a graded algebra. We have not given a name to this algebra or introduced a special notation for it though we have shown

(Chapter 7, Theorem 16) that it is isomorphic to the algebra of differential operators on $S(M)$. However, the proof of Theorem 1 shows that when M is a finite free module the algebra in question is the graded dual of the coalgebra $S(M)$; hence there is no harm in using $S(M)^\dagger$ to denote the algebra in question in all cases. Thus, to sum up, *$S(M)^\dagger$ is defined for an arbitrary module M, and when $S(M)$ has finite free components it coincides with the graded dual of the coalgebra $S(M)$.* Note that $S(M)^\dagger$ is always a commutative algebra and that the module formed by its homogeneous elements of degree one is none other than M^*.

Because $S(M)^\dagger$ is commutative, the inclusion mapping $M^* \to S(M)^\dagger$ extends to a homomorphism (of algebras)

$$\lambda: S(M^*) \to S(M)^\dagger \tag{8.5.5}$$

which in degree p $(p \geq 1)$ induces an R-linear mapping

$$\lambda_p: S_p(M^*) \to S_p(M)^*. \tag{8.5.6}$$

The way in which (8.5.6) operates is described in the next exercise.

Exercise 3. *Let $f_1, f_2, \ldots, f_p$ $(p \geq 1)$ belong to M^*. Show that $\lambda_p(f_1 f_2 \ldots f_p)$ is the linear form on $S_p(M)$ that maps $m_1 m_2 \ldots m_p$ into the permanent of the matrix $\| f_i(m_j) \|$.*

The permanent of a square matrix was defined in Section (6.8).

Now suppose that M is a finite free module. Let $b_1, b_2, \ldots, b_s$ be a base of M and $f_1, f_2, \ldots, f_s$ the base of M^* that is dual to it. Then the elements $b_1^{\mu_1} b_2^{\mu_2} \ldots b_s^{\mu_s}$, where $\mu_i \geq 0$ and $\mu_1 + \mu_2 + \cdots + \mu_s = p$, form a base of $S_p(M)$; and likewise the elements $f_1^{\nu_1} f_2^{\nu_2} \ldots f_s^{\nu_s}$, where $\nu_i \geq 0$ and $\nu_1 + \nu_2 + \cdots + \nu_s = p$, form a base for $S_p(M^*)$.

Exercise 4. *Show that*

$$\lambda_p(f_1^{\nu_1} f_2^{\nu_2} \ldots f_s^{\nu_s})(b_1^{\mu_1} b_2^{\mu_2} \ldots b_s^{\mu_s})$$

has the value $(\nu_1 !)(\nu_2 !) \ldots (\nu_s !) 1_R$ if $(\nu_1, \nu_2, \ldots, \nu_s) = (\mu_1, \mu_2, \ldots, \mu_s)$ and is zero otherwise.

Exercise 4 shows that it is possible for $\lambda_p(f_1^{\nu_1} f_2^{\nu_2} \ldots f_s^{\nu_s})$ to be zero. Consequently, when M is a finite free module the homomorphism $\lambda_p: S_p(M^*) \to S_p(M)^*$ need not be an isomorphism and therefore the algebra-homomorphism $\lambda: S(M^*) \to S(M)^\dagger$ need not be an isomorphism either. However, in this situation $S(M^*)$ and $S(M)^\dagger$ are certainly graded Hopf algebras.

Exercise 5. *If M is a finite free module show that the mapping $\lambda: S(M^*) \to S(M)^\dagger$, defined in (8.5.5), is a homomorphism of graded Hopf algebras.*

This can be solved by arguments very similar to those used in connection with Exercise 2.

8.6 Solutions to selected exercises

Exercise 1. *Let the notation be as explained. Show that the base* $\{f_J\}$ *of* $E_p(M^*)$ *is mapped by* (8.5.3) *into the base of* $E_p(M)^*$ *that is dual to the base* $\{b_I\}$ *of* $E_p(M)$.

Solution. Suppose that (8.5.3) takes f_J into the linear form ϕ_J. Then, by (8.5.4),

$$\phi_J(b_I)=\begin{vmatrix} f_{j_1}(b_{i_1}) & f_{j_1}(b_{i_2}) & \cdots & f_{j_1}(b_{i_p}) \\ f_{j_2}(b_{i_1}) & f_{j_2}(b_{i_2}) & \cdots & f_{j_2}(b_{i_p}) \\ \vdots & \vdots & & \vdots \\ f_{j_p}(b_{i_1}) & f_{j_p}(b_{i_2}) & \cdots & f_{j_p}(b_{i_p}) \end{vmatrix}.$$

Now if $(i_1, i_2, \ldots, i_p)\neq(j_1, j_2, \ldots, j_p)$, then we can find t so that j_t is not in I and hence $\phi_J(b_I)=0$ because the determinant has a row of zeros. On the other hand, if $I=J$, then $\phi_J(b_I)=1$.

Exercise 2. *If* M *is a finite free module, show that the mapping* $E(M^*)\to E(M)^\dagger$ *defined above is an isomorphism of modified Hopf algebras.*

Solution. We already know that $E(M^*)\to E(M)^\dagger$ is an isomorphism of graded algebras so it will suffice to show that it is also a homomorphism of coalgebras.

Let us denote the homomorphism under consideration by θ and for g in $E(M^*)$ use g^θ to denote its image in $E(M)^\dagger$. To show that θ is compatible with comultiplication we need only show that

$$\begin{array}{ccc} E(M^*) & \xrightarrow{\Delta_1} & E(M^*) \underline{\otimes} E(M^*) \\ \Big\downarrow{\scriptstyle\theta} & & \Big\downarrow{\scriptstyle\theta\otimes\theta} \\ E(M)^\dagger & \xrightarrow{\Delta_2} & E(M)^\dagger \underline{\otimes} E(M)^\dagger \end{array}$$

is a commutative diagram, where Δ_1 and Δ_2 are comultiplications. But all the mappings are homomorphisms of algebras and M^* generates $E(M^*)$ as an algebra. Thus we need only show that $\Delta_2(f^\theta)=(\theta\otimes\theta)(\Delta_1(f))$ for all f in M^*. Now

$$(\theta\otimes\theta)(\Delta_1(f))=(\theta\otimes\theta)(f\otimes 1+1\otimes f)=f^\theta\otimes\varepsilon+\varepsilon\otimes f^\theta,$$

where ε is the identity of $E(M)^\dagger$, that is to say it is the projection of $E(M)$ onto $E_0(M)=R$.

Let μ be the multiplication mapping of the algebra $E(M)$. Then Δ_2 is obtained by combining the homomorphisms

$$E(M)^{\dagger} \xrightarrow{\mu^{\dagger}} (E(M) \otimes E(M))^{\dagger} \xrightarrow{\sim} E(M)^{\dagger} \otimes E(M)^{\dagger}.$$

Also $\mu^{\dagger}(f^{\theta}) = f^{\theta} \circ \mu$. Consequently it suffices to show that the isomorphism $(E(M) \otimes E(M))^{\dagger} \approx E(M)^{\dagger} \otimes E(M)^{\dagger}$ matches $f^{\theta} \circ \mu$ with $f^{\theta} \otimes \varepsilon + \varepsilon \otimes f^{\theta}$ and for this it is enough to establish that

$$f^{\theta}(x \wedge y) = \varepsilon(y) f^{\theta}(x) + \varepsilon(x) f^{\theta}(y) \tag{8.6.1}$$

for all homogeneous elements x and y of $E(M)$. Now, except in the case where one of the elements has degree zero and the other has degree one, both sides of (8.6.1) are zero; and if we do have this special situation, the two sides are obviously equal. Thus θ is compatible with comultiplication.

It remains to be shown that θ also preserves counits, that is to say we must show that the diagram

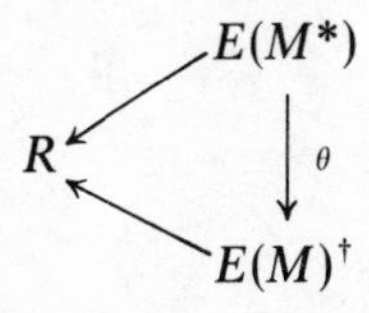

is commutative, where the unlabelled mappings are counits. But, once again, the mappings are homomorphisms of algebras, so that we need only examine what happens to an element of M^*. However, for such an element its image in R is zero by either route. Accordingly the solution is complete.

Index